AIDS TO
GEOGRAPHICAL
RESEARCH

AIDS TO GEOGRAPHICAL RESEARCH

*Bibliographies, Periodicals, Atlases, Gazetteers
and Other Reference Books*

By
JOHN KIRTLAND WRIGHT
Director, American Geographical Society
and
The Late ELIZABETH T. PLATT
Librarian, American Geographical Society, 1937–1943

SECOND EDITION
COMPLETELY REVISED

GREENWOOD PRESS, PUBLISHERS
WESTPORT, CONNECTICUT

ACKNOWLEDGEMENTS

THE FIRST edition of this book, by the undersigned, appeared in 1923. It carried the subtitle "Bibliographies and Periodicals" and constituted No. 10 of the *American Geographical Society Research Series*. The present edition has been completely revised not only with a view to bringing the first edition up to date but to include, in addition, atlases and certain other geographical reference books. A wholly new Introduction has also been provided. Nearly all of the references to bibliographies and periodicals that had appeared prior to 1942 and were not already listed in the first edition were accumulated by Elizabeth T. Platt. At the time of her tragic death in May, 1943, we were planning to work together on the volume with the thought that she would do the greater part of the work. Had such collaboration been possible this would have been a better book.

Grateful acknowledgement is made of the assistance which the following persons rendered in preparing the manuscript for the press: Mrs. Luella Dambaugh, formerly of the staff of the Society; Miss Alice Taylor, Mrs. Wilma B. Fairchild, Miss Marion Lack, the late Miss Elsa Rowell, and others of the staff of the American Geographical Society; Mr. Saul Benison, graduate student in the Department of History, Columbia University, and Miss Alice E. Pratt of the Columbia University Press. The help of 210 American geographers in preparing the Appendix is acknowledged on page 278, below.

JOHN K. WRIGHT

New York, April 10, 1947

CONTENTS

II: Topical Aids

III: Regional Aids and General
Geographical Periodicals

INTRODUCTION

INTRODUCTION

THIS BOOK is intended to be of help to anyone who has occasion to make a serious study of geography. It is designed more particularly to serve advanced students and professional workers in that field in carrying on research in libraries and in preparing manuscripts for publication. In the Introduction some observations on the nature of geographical studies are followed by a general discussion of published aids to geographical research. The main part of the volume consists of selective lists of bibliographies, periodicals, atlases, and other reference works. The purpose is not to furnish the student with references to the primary works on which to base his investigations, for no single book could do this adequately for the whole immense subject of geography. The reader will look in vain for the titles of publications on, say, the geography of plants, or of cities, or of Massachusetts. He will, however, find information guiding him to other works, geographical or otherwise, which should, in turn, open up the literature and the maps pertaining to almost any aspect of geography that he may wish to study.

THE NATURE OF GEOGRAPHICAL STUDIES

Motives for Geographical Investigation

To survive on this planet, the higher animals acquire geographical knowledge of a rudimentary kind. They must know something of the distribution of things on the lands and in the seas that form their habitats—where to go to find food or shelter or comfort, or to escape from dangers. If instinct rather than intelligence or volition controls their movements, instinct itself springs from a mysterious kind of knowledge: the migratory birds know where to fly over thousands of miles of continent and ocean, though their ways of knowing differ materially from the geographer's. The most primitive of peoples also are familiar with the distribution of resources and obstacles in terrains where they fish, hunt, and fight, and many travelers have told of the remarkably accurate and detailed maps that wandering tribesmen often carry in their minds' eyes and can scratch on turf, or sand, or snow. Among more advanced folk such knowledge is widened, deepened, systematized, and, since there are few enterprises necessary to modern human existence or progress in which geographical circumstances do not have to be reckoned with, innumerable persons are today engaged either directly or indirectly in the gathering, interpreting, and disseminating of geographical information.

We may distinguish three motives for the advancement of geographical knowledge: personal, group, and disinterested. Personal motives impel the individual to carry on geographical investigation for his own benefit or satisfaction; group motives impel him to do so in the interests of groups of people; and disinterested motives impel him to do so in the interests of humanity.

Personal motives.—Many individuals today embark upon geographical studies because their jobs so require. Some, however, go into or continue in geographical work because there is something about it that allures them. These are geographers by nature, gifted in geography as others are gifted in music, painting, or mathematics. But while an inner fire may give zest to studies in the field or library, the satisfaction of social needs rather than of personal desires is the main incentive to geographical investigation.

Group motives.—Man conducts most enterprises necessary for his survival and improvement on a coöperative basis. The organized groups by which geographical research is conducted fall into two categories: (1) professional or amateur groups united by an interest in geography as such; and (2) groups united by other interests, for the promotion of which geographical knowledge is necessary.

The groups of the first kind include societies of geographers and of laymen who are interested in geography; associations of professional geographers and of teachers of geography; explorers', travelers', and mountaineering clubs; departments and institutes of geography in universities and other educational institutions; and a multitude of scientific societies, museums, and research organizations, which specialize in such subjects as geology, climatology, oceanography, biology, ecology, anthropology, and the other natural sciences, and in the social studies. These organizations may be local, national, or international in scope. Many issue publications and a few maintain research staffs. While the activities of some of these groups are designed first and foremost to benefit their own members, most of the groups seek to serve broader social purposes. Frequently groups in this category undertake geographical investigations in the interests of groups of the second category, and their services are enlisted by governments, especially in wartime.

Groups united by interests that are not geographical, as such, are of three main kinds: religious, economic, and political.

Religious groups, as represented by the church and the great monastic orders, once played a vigorous part in the advancement of geography. Much of the geographical learning of the Middle Ages reflected attempts of clerical scholars to reconcile lore handed down from classical antiquity and information derived from firsthand observations with the Bible and the teachings of the Church Fathers. Much more substantial contributions to geography, however, have been made by the Christian missionaries—in the great age of discovery by Dominicans, Franciscans, and Jesuits, and more recently by Protestants.

Among the economic groups which gather geographical information, we may refer especially to commercial, industrial, and public-utilities companies. Geography owes an immense debt to the great English and Dutch trading companies of the seventeenth and eighteenth centuries, and later to the whaling, rubber, mining, and oil companies whose operations in little-known and undeveloped parts of the world have necessitated the making of countless surveys, maps, and geographical reports. Modern corporations serving highly developed communities also undertake detailed surveys of the distribution of population and resources, and of social and economic conditions in the areas where their products or services are sold.

Political groups, represented especially by national governments but also by the governments of colonies, states, provinces, counties, cities, and the like, probably do more than all other organized groups to promote the advancement of geography. They maintain agencies which carry on surveys in their own territories and, in the case of many sovereign states, in other parts of the world as well. Most of the detailed and systematic information today available concerning the topography of the lands, the hydrography of the oceans, and the distribution of natural resources, climatic elements, population, industries, occupations, and social and economic conditions is derived from maps, statistics, and other documents issued by governments.

In conducting surveys and other geographical studies governments usually have in view either the publication of information for their citizens to use in private enterprises, or the securing of information—some of which may not be published—primarily for the use of governmental agencies. Many studies serve both purposes.

The interests of the professional group to which a geographer belongs—whether it be university, government agency, or research institution—influence the choice of the subjects of his investigation, although he often enjoys considerable freedom. Beyond the immediate group, the intellectual, social, political, and economic climates of the place and time in which he lives provide the conditions of his personal preference. These changing climates favor some types of study and blight others. Few geographers have the hardihood and opportunity deliberately to pursue studies to which they know the climate is hostile, though to do so might on occasion be in the best interests of science and of society.

These climates have been disturbed since 1914. The storms of war, economic depression, and human suffering have profoundly affected the trends and the objectives sought in geographical investigations of all kinds. They have placed a premium of social approval on research and educational programs that bear directly on large problems of national, regional, or local welfare, as distinguished from studies of a more recondite kind. In Germany the storms produced Geopolitik. In the United States, they have led to a closer collaboration between geographers and scholars in other fields in the coöperative investigation of many problems—such as those connected with the exploitation and conservation of natural resources, the movements of peoples and the opening up of undeveloped regions for new settlement, and a host of problems related to the winning of the war, and today to the establishment of peace. The tabulation given on pp. 276–294, below, of American geographers according to the subjects of their research sheds light upon these problems.

Many of these problems are concerned with plans of action. The geographer's function here is to help provide the foundations of integrated data upon which action must be based and to which it must be adjusted if it is to succeed. Frequently such research is conducted largely with a view to its potential value in future action along lines not definitely foreseen. Sometimes, on the other hand, it is carried out as a fundamental part of the formulation of a plan—whether for a village, a city, a region, a nation, or for the reconstruction of a war-shattered world.

Disinterested motives.—The noblest motives that impel individuals to carry on geographical studies—or any other kind of investigation, for that matter—are those that are disinterested, that spring from a desire to serve mankind as a whole through the objective quest for truth. Practices of two main kinds impede this quest. One is the waste of time, energy, and talent upon studies that may yield personal satisfaction but add little that is new or of consequence. The other, more serious, is the suppression or distortion of the truth as a result of conflicting individual or group pressures.

Obviously no prescriptions can be laid down to guide one in deciding whether the geographical investigation of a particular topic or area is worth while. It might well be borne in mind, however, that, if the geographer wishes to select a small area for study, he should make sure that no one else has already dealt in much the same manner with an essentially similar area in the same region. Otherwise his work, though it might be of some local interest to the inhabitants of the area itself, will add nothing substantial to human understanding. There are too few geographers to warrant research of extremely narrow scope unless it discloses principles of wider application, or to justify reduplication in purpose and ideas or the adding of mere embroidery to basic concepts already developed.

If new truths are to be learned, men must share the knowledge that they have already acquired. The advancement of science is retarded by the failure to publish or otherwise make freely available the results of investigations. Human fear, jealousy, and selfishness cause many individuals and groups to keep their knowledge to themselves. Even scholars and scientific institutions are sometimes secretive about researches which they have under way lest others steal the results and deprive them of the kudos they might otherwise receive. Temporary secretiveness is often justified, but when prolonged, it has been known to degenerate into a dog-in-the-manger attitude. Business corporations which carry on geographical studies in their own interests are more likely than not to keep their maps and reports locked in confidential files. Among government agencies bureaucratic policy frequently brings about the suppression of the results of important studies through fear of criticism in the press, in Congress, and from higher government officials. National interests demand the suppression of geographical information of value to actual and potential enemies, and during wartime the dissemination of much geographical information—particularly in the form of detailed maps and statistics—is subject to a complete scientific blackout in so far as the public is concerned. Nearly every geographer sooner or later becomes acutely aware of the fact that, in addition to the data that he may observe in the field and gather freely in the library, there also exist large stores of information concerning matters on which he is working to which access is difficult and often impossible. Tact, diplomacy, "pull," and, best of all, a reputation for devotion to disinterested scholarship frequently may overcome many of the obstacles.

FIELD WORK AND INDOOR WORK

The primary facts with which geographical research has to do cannot be brought into a laboratory and examined under a microscope or in a test tube. They may be directly observed in the field only. It does not follow, however, that no geographical study is scientifically sound unless it is undertaken in the field. Most geographical investigations deal with subjects far too complex and regions much too large for any human being personally to observe and record all that is needed. Reliance must be placed, in part at least, on the records of firsthand observations that others have made.

Geographical investigations in which the observations of others are accepted uncritically are of course unscientific. Much of the geography of the learned writers of the Middle Ages was of this nature—a body of erudition with little scientific quality in the modern sense despite the diligence of the compilation and the ingenuity of the reasoning that it often discloses. Many of the matters that the modern geographer has occasion to investigate *must* be studied in the field. This is particularly true of problems that affect the welfare of specific communities in specific areas—problems on which existing maps, statistics, and reports shed little or no light and about which the most revealing data can be discovered only by looking for them on the spot, talking with people, examining the earth's face for telltale evidence that others have missed, and making one's own new maps and new statistics. Furthermore, field studies normally add life and authenticity to geographical writings—though many geographical works of distinction have been based on "indoor" work alone.

It should be borne in mind that in some respects geography is more akin to history—which is based almost entirely on the interpretation of documentary evidence—than it is to physics or chemistry or geology. But even the physicist or chemist must go to the library and take stock of what others have done and are doing in his subject.

In short, the advancement of geography demands both field work and the examination of documents, and it hardly behooves either the more "bookish" or the more "out-of-doors" geographer to rationalize a personal preference for the one or the other mode of investigation into a conviction that it is intrinsically the more worthy from the scientific point of view.

GEOGRAPHICAL DESCRIPTION AND INTERPRETATION

The immediate purpose of geographical research is the description or interpretation, or both, of certain terrestrial facts. *Geographical description* discloses or portrays the relationships in space on the earth's surface of different facts to one another, as the relationship, for example, between the position of New York and the course of the Hudson River. *Geographical interpretation* aims to show *why* such relationships exist or have existed.

Many geographical works consist entirely, or in large part, of descriptive matter: for example, gazetteers, census statistics, geographical handbooks, books of travel, lectures of the travelogue type. The data presented on the majority of maps are descriptive rather than explanatory, as is also the information that statistical analysis may yield with regard to mathematical correlations between the distributions of different phenomena. Description is often based on extremely refined methods of analysis and synthesis, requiring technical skill and judgment of a high order. Accurate description is the foundation for interpretation. If our information as to how things are arranged on the earth's surface is unsound—if it is based, for example, on faulty generalizations or uncritical acceptance of information at second or third hand—our conclusions as to why they are so arranged are likely to be shaky.

It is possible to draw inferences as to cause and effect directly from descriptive geographical works—maps, for example. One may cor-

rectly conclude from a map that a certain region is uninhabited because it is a desert or that a bend in a railroad is due to the windings
in a valley, or that a town is located in a particular place because there
is a coal field nearby. While these are not the sole causes—there is no
sole cause for anything—they are at least important causes among
many. The map, however, does not state them as causes, but leaves it
to the reader to draw his own conclusions. In their professional studies, geographers are seldom content with merely describing terrestrial
relationships or portraying them as does a map, and letting it go at
that. One of the main objectives in geographical research is the discovery and understanding of causal connections between terrestrial
circumstances. Geographical interpretation seeks to disclose such connections.

Many of these connections defy ready explanation in terms of what
now exists. Why does a river run parallel to a range of mountains for
fifty miles and then turn abruptly and cut through the range in a
deep gap? Why is a port located on a particular harbor, when there
are equally good harbors in the vicinity? To explain relationships of
this sort one must look back into the past—either to the geologic past
or the human past. Hence, in so far as geographical research seeks to
explain and interpret, the element of time must be reckoned with—
just as, conversely, it is inescapable that the element of space must be
reckoned with in historical research. For this reason history and inter-
pretative geography are inseparable, however much their purposes
may differ.

The ultimate purpose for which a geographical study is undertaken
and the good judgment of the investigator determine the particular
causal connections sought for and how far chains of causation need
be traced back in time. Frequently a general cause for the existence of
geographical circumstances may be discovered but not the specific
cause at a particular place. We may, for example, explain the presence
of landslides on a mountain range with reference to slope, rock, soil,
and climate; but the occurrence of a particular slide at a particular

place may be due to some unknowable and accidental event—a stroke of lightning, perhaps. Many of the facts of "human geography" today can be explained only by reference to unrecorded happenings, some of them far from the places where their effects are now evident.

Geographers have long been especially concerned with the causal connections between "man" and "nature." Here the processes of cause and effect operate in two directions. Man compels nature to change its aspects when he clears forests, reclaims swamps, dams streams. Nature, on the other hand, compels or induces man to do certain things and acquire certain characteristics. It is not always easy to draw clean-cut distinctions between these processes. When a flood sweeps away a house and family there is active physical compulsion on the part of nature; and nature alters men's bodies through the direct physiological influences of climate, insects, bacteria, and the like. Most human activities and conditions, however, are determined immediately and directly by the human mind and only indirectly by nature. When men leave a countryside because of drought, the drought does not physically push them out; they leave, rather, because they fear the consequences of staying, and such an exodus might therefore be called an enforced adaptation to nature. When men build roads along valleys instead of over hills, or cultivate fertile soils and leave the barren lands unoccupied, they both alter nature and voluntarily adapt themselves to it. Because it is hard to draw the line between the acts that are voluntary and those that are enforced, it is often a question of playing with words whether a human condition be called the result of environmental control or merely a voluntary adaptation to the natural environment. Yet geographers have argued at length over this matter. The tendency today among American geographers is toward caution in explaining human activities and conditions as the direct result of compulsion by nature. "Environmental determinism" is not so common as it was when the complexity of geographical circumstances and their causes was less fully appreciated. Most of our geographers are circumspect in generalization with regard to "influences of the natural environment"—more so, perhaps, than some of their colleagues in other fields.

Through their professional studies geographers have done much to increase and deepen knowledge of what nature impels man to do and be and of what man does to nature. The soundest work of this kind has dealt with specific conditions in specific regions. The pressure of social and national interests is leading to increasing emphasis on research designed to serve as a basis for definite prescriptions with regard to better adaptations that man may make both to and of nature.

In discussions of geographical research the idea is occasionally expressed, or implied, that geographical description is of a lower order of merit than interpretation. The fallacy here is not unlike the fallacy that field work is intrinsically more "scientific" than library work. Whether or not, in a particular study, a geographer should engage primarily in explaining or primarily in describing depends on the purpose of the study. If the purpose, for instance, be to present information basic to a plan of action—a military campaign, the settlement of a region, the future development of a city—description may be far more to the point than explanation. Similarly, in a study undertaken not for any immediate practical reason but in order to give a more or less well-rounded understanding of a region as a whole, it may not be at all desirable to confine the treatment to matters capable of explanation and to the explanation of those matters, since many of the spatial relationships that give the region its character either defy explanation or else may be adequately explained only by going very far afield. In other words, judgment and balance play a large part in geographical research, which is an art for which specific definitions and rules cannot be laid down; an art in which the media and methods must be adapted to the particular problems that have to be solved; but, nevertheless, an art that must always be guided by a scientific habit of thought.

AIDS TO GEOGRAPHICAL RESEARCH

The Subject Matter of This Book

An aid to geographical research is anything actually used to facilitate research in geography, be it a published or unpublished document, a surveying instrument, a word of helpful advice, a sum of money. This book is concerned almost exclusively with published documents.

There are innumerable ways of classifying documents. A fundamental distinction is between classifications according to the subjects with which the documents deal and classifications according to other criteria. In this volume the references are entered under main headings representing their subjects and under subordinate headings representing other characteristics (for example, whether or not they are periodicals or government publications).

The material in this volume is classified primarily according to *geographical subjects.* Many aids to geographical research, however, are not specifically or exclusively geographical in nature or purpose. Some are designed for use in connection with investigations in fields allied to geography, such as history or geology. Others are universal or broadly comprehensive aids to investigations in all or many different fields, including geography.

Grouped according to their geographical subjects, aids to geographical research fall into three main categories:

1. There are aids of which the geographical subject in each case, either embraces the entire field of geography or at least is not definitely restricted to any specified parts of that field. Such aids, in turn, are of two kinds:

a) Universal or comprehensive aids to studies in many different fields, including geography. References to these are given in the Introduction, though a few are also listed in Part I.

b) General geographical aids. References to these are given in Part

I, General Aids, and a few are also listed in the Introduction. General geographical periodicals, however, are listed in Part III.

Since the aids in category 1 all bear potentially on the whole field of geography, they may be subclassified according to nongeographical criteria only. In the Introduction and in Part I a primary distinction is made between bibliographical and nonbibliographical aids.

2. There are aids of which the geographical subject in each case, though unrestricted in scope to any specified region, is nevertheless limited to the whole or a part of one of the main topical divisions of geography (for example, physical geography, geomorphology, human geography, city geography).[1] References to aids of this type are given in Part II, Topical Aids, under headings representing various topics. Aids of this category are also of two kinds:

a) Nongeographical aids bearing on fields allied to the several divisions of geography (for example, history, allied to historical geography; geology, allied to geomorphology). A number of these of very broad scope are listed in Part II.

b) Geographical aids bearing specifically on the several topical divisions and subdivisions of geography as such.

3. There are aids of which the subject in each case is limited in scope to a specified region or group of regions. These, together with general geographical periodicals, are listed in Part III, Regional Aids, under regional headings. The regional aids pertaining to any given region or subregion also fall into four main groups corresponding to the groups enumerated under 1 and 2.

Most of the references in this book are to bibliographical aids— works of which the purpose is to provide information about other works. Certain nonbibliographical aids, however, also are listed. These comprise a selection of world and regional atlases and also a

---

[1] American geographers customarily designate these as "systematic" divisions. In this book, however, the term "topical" is consistently used with reference to classification and "systematic" only with reference to a logical arrangement of classes or of items within each class. See below, pp. 31, 49.

number of comprehensive reference books such as manuals of general and regional geography, world gazetteers, and guides to the spelling of place names. Geographical periodicals and other geographical serials are also included, some of them of bibliographical value, others nonbibliographical.

Geographical as well as universal and comprehensive aids will be discussed in general terms in this Introduction. With a few exceptions, however, the references given in the text and footnotes of the Introduction are exclusively to universal aids and to certain comprehensive aids to research in fields allied to geography.

The reader will find that certain aids of outstanding utility not only are listed by title in the Bibliography but also are cited briefly with page references under various topical and regional headings. Works treated in this manner are the sixth edition of Mudge's *Guide to Reference Books* [1],[2] Lewin's *Subject Catalogue of the Library of the Royal Empire Society* [337], Martineau's *Bibliographie d'histoire coloniale* [331], and a few others. The titles of the most recent reports on the progress of geographical research appearing in the comprehensive *Geographisches Jahrbuch* [6] have also been entered under the appropriate headings, and care has been taken to include references to recent materials of bibliographical interest in the *Geographical Review* [476], a periodical readily available to American students. Owing to limitations of space, however, other geographical periodicals have not been analyzed to the same extent, nor do the two leading current geographical bibliographies, *Bibliographie géographique internationale* [7] and *Current Geographical Publications* [9] lend themselves to such analysis.

THE MUDGE-WINCHELL GUIDE

One bibliographical aid that might well be constantly used in conjunction with the present book deserves special mention at this point.

―――――――――――

[2] Figures in square brackets refer to the serial numbers of the references in the Bibliography, below.

This is the sixth edition of Mudge's *Guide to Reference Books* with its three supplements (cited herein as "Mudge-Winchell" [1]). When in search of initial guidance to publications in almost any field, the American student would do well to consult this manual, which should be available in every library of any consequence in the United States.

Mudge-Winchell is divided in two main parts, of which the first deals with reference books of universal scope and the second with those pertaining to "special subjects." In the first part are listed encyclopedias, dictionaries, and other works having to do with periodicals, essays, and general literature, debates, dissertations, and the like. The special subjects include the social sciences, science (that is, mathematics, the natural sciences, and ethnology), the useful arts, the fine arts, literature, biography, geography (confined mainly to gazetteers, geographical names, atlases, and guidebooks), history, government documents, and bibliography.

GUIDES TO INDIVIDUALS AND INSTITUTIONS

Although the present volume deals mainly with publications, the fact should not be overlooked that people and institutions may also be used effectively as aids to geographical research. Time can often be saved by consulting geographers, explorers, or librarians, as long as their good nature is not imposed on unduly. Geographical institutions such as the Association of American Geographers,[3] the American Society for Professional Geographers,[4] the Division of Geology and Geography of the National Research Council,[5] and the geographical societies can usually put the student in touch with individuals familiar with particular regions or branches of geography, and a classified list of such individuals will also be found in the Appendix, pp. 276–294, below. During the war there was established in Washington the comprehensive National Roster of Scientific and Spe-

[3] Professor Chauncy D. Harris, Secretary, Department of Geography, University of Chicago, Chicago, Illinois.
[4] Dr. E. W. Miller, Secretary, Pennsylvania State College, State College, Pa.
[5] Miss Margaret L. Johnson, Secretary, 2101 Constitution Avenue, Washington, D.C.

cialized Personnel, an immense classified card index of experts in many fields. Special lists of geographers, cartographers, and travelers compiled by the geographical division of the Office of Strategic Services have been deposited in the National Archives, Washington. The Ethnogeographic Board, with headquarters at the Smithsonian Institution, has compiled a card index of persons throughout the world "having varying degrees of familiarity with one or more foreign areas."

Biographical reference works on living persons—the various *Who's Who, American Men of Science,* and the like—are also of service.[6] Naturally, examination of recent publications on a given topic or region will often acquaint one with the names of persons who might be consulted.

In planning almost any project of geographical research the student will do well to familiarize himself with the scientific organizations that are or have been concerned with the investigation of the topics or areas in which he is interested. Not only is it likely that their publications will be of value to him, but such organizations can often give him active help by permitting him to use their facilities and by recommending him to others. Before the war a number of guides to the learned institutions of all nations were published from time to time, the best known and most comprehensive of these being *Minerva: Jahrbuch der gelehrten Welt* [7] and *Index Generalis.* [8] In these may be

---

[6] Mudge-Winchell [1], a 281–314, b 39–45, c 57–65, d 55–59. See also *Biographical Sources for Foreign Countries* (The Library of Congress, General Reference and Bibliography Division, Washington, mimeographed). 1, *General* (compiled by Helen D. Jones), 1944; 80 pp. 2, *Germany and Austria* (compiled by N. R. Burr), 1945, 215 pp. Further numbers were planned to deal with the remainder of Continental Europe, the U.S.S.R., the British Commonwealth of Nations, and Asia.

[7] *Minerva: Jahrbuch der gelehrten Welt,* ed. by Gerhard Lüdtke (Berlin and Leipzig, W. de Gruyter & Co., 1891–1914, 1920–[1937–1938]. The most recent edition is in two parts: Part 1, 1937, covers research institutions, learned societies, observatories, libraries, archives, museums, etc.; Part 2, 1938 (2 vols.), covers universities and technical high schools (*Fachhochschulen*). Concise information is given concerning each institution with the names of its principal administrative and scholarly personnel. In each part the material is arranged alphabetically by towns, and each part has several indexes.

[8] *Index Generalis: The Year-Book of the Universities, Colleges, Schools of Science and*

found entries for universities, libraries, academies, archives, scholarly and scientific societies, museums, observatories, and the like, with data on their objectives, collections, and publications, and lists of their administrative, educational, and research personnel. International organizations are listed and their activities briefly described in a handbook issued by the League of Nations in 1938 [9] and information with regard to the publications of international congresses and conferences will be found in a union list [10] which appeared in the same year. For a more extended introduction to the bibliography of such organizations, Mudge-Winchell should be consulted.[11]

Aids to Geographical Field Research

In going into the field the geographer is confronted with problems both of travel and of investigation. The purely practical problems of travel—organization, costs, equipment, leadership, protection from climate and disease, may be easy or difficult to solve, depending on the accessibility of the region visited and the size and purpose of the expedition. There are a number of manuals of advice for travelers [86, 87]. Scattered information may also be found in the published reports of scientific expeditions. Usually, however, published information is not altogether satisfactory, and the best policy in planning field

Technology, Astronomical Observatories, Scientific Institutions, Libraries, Learned Societies, issued under the direction of R. de Montessus de Ballore (Paris, 1919–[1939]), annual.

[9] League of Nations, *Handbook of International Organizations (Associations, Bureaux, Committees, etc.)* (Geneva, 1938), 491 pp.

[10] Winifred Gregory, *International Congresses and Conferences, 1840–1937: A Union List of Their Publications Available in Libraries of the United States and Canada* (Bibliographical Society of America) (New York, H. W. Wilson, 1938), 231 pp. Excludes diplomatic congresses and conferences, and those held under the auspices of the League of Nations.

[11] Mudge-Winchell [1], a 34–38, c 13–14, d 11–12. See also Harriet W. Pierson, *Guide to the Cataloguing of the Serial Publications of Societies and Institutions* (Library of Congress), 2d ed. (Washington, Government Printing Office, 1931), 128 pp. In the section entitled "Bibliographical Suggestions," pp. 76–112, indication is given of "the kinds of sources in which information concerning societies and institutions may be found."

work in a particular area, if you have not been there before, is to consult others who have and profit from their counsel. The war, however, has given an immense impetus to studies of the needed equipment, supplies, and techniques for the conduct of expeditions in all parts of the world. Especially noteworthy investigations have been made by the United States Quartermaster Corps and the Arctic, Desert, and Tropic Information Center of the United States Army Air Forces. Some of the information that was accumulated by the armed forces during the war is now being made available to the public.

Techniques of field research for certain kinds of geographical investigation have been standardized and in many instances are explained in manuals and textbooks. This is true of comprehensive surveys of large areas conducted by governments and corporations—for example, geodetic,[12] topographic,[13] hydrographic,[14] or geological [15] surveys which require instrumental observations. It is especially desirable that the professional geographer have some acquaintance with at least the elementary principles of topographic mapping in order that he may make his own base maps where suitable published maps are not available for use as bases.[16]

Unlike the surveyor or census taker, however, the professional geographer in his field studies can seldom follow any standardized rules as to what to observe and how to record his observations. Although he may sometimes find that techniques already developed by others are appropriate, more often he will be obliged to devise his own

[12] William Mussetter, *Manual of Reconnaissance for Triangulation* (Washington, 1941) (*U.S. Coast and Geodetic Survey Special Publ.*, No. 225).

[13] H. M. Frye and others, *Topographic Instructions of the United States Geological Survey* (Washington, 1928) (*U.S. Geol. Survey Bul.* 788, 1928).

[14] K. T. Adams, *Hydrographic Manual* (Washington, 1942) (*U.S. Coast and Geodetic Survey Special Publ.*, No. 143).

[15] F. H. Lahee, *Field Geology*, 4th ed. (New York, McGraw-Hill, 1941), 853 pp.

[16] A. R. Hinks, *Maps and Survey*, 4th ed. (Cambridge Univ. Press, 1942), 313 pp. Designed especially for students of geography and by the late Secretary of the Royal Geographical Society, this book contains a chapter on "Exploratory Survey," in which relatively simple methods of field map sketching are set forth.

methods. Specific field methods that have been followed by professional geographers have been described and discussed in countless monographs and periodical articles, but this material has not been gathered together and presented systematically in a single work. A comprehensive manual on the subject would be useful indeed, even though it did little more than give well-selected references to publications.

Aids to Geographical Study Indoors

The source materials upon which indoor geographical studies are based consist both of man-made objects, such as tools, artifacts, books, maps, and other documents, and of natural objects, such as specimens of minerals, soils, plants, animals. These materials are housed in libraries, archives, museums, and private collections.

Among the man-made objects employed in indoor geographical research, certain ones have been designed for the express purpose of conveying geographical information. These alone are properly "geographical documents," though, obviously, geographical information of great value may be derived from other documents and materials. A novel or a poem may convey such information. A piece of pottery when studied in relation to other artifacts may shed indirect light on the geographical distribution of a human culture. Hence aids to investigations of many different kinds may also serve as "aids to geographical research."

Methods of geographical research based on written and printed documents are so varied that no one has attempted to compile a general compendium of such methods. All documentary research, however, necessitates in the first instance critical analysis of the documents themselves. In his field studies the geographer is directly face to face with geographical realities, but when he uses documents that were designed to convey information and ideas about these realities, he sees the realities only as they have already been pictured in other men's minds. Consequently he must be doubly on guard against his own intellectual limitations and those of the minds that intervene between him and the realities. While his interpretation of what the documents actually say may be above reproach, his whole study will be founded on sand if he cannot distinguish the true from the false or the suspect in the documents themselves. Procedures for the critical analysis of written documents have been developed and systematized by historians, and geographers might do well to enlarge their acquaintance with some of the historians' works on this subject.[17] Students of the social sciences, on the other hand, are often more prone than are geographers to take geographical documents at their face value and might profit by greater awareness of the possibilities of error in the evidence that such documents—maps, especially [18]— present.

Guides to collections.—The source materials used in indoor research have been assembled in collections of two kinds, "closed" and "open." Closed collections are those that are housed in private dwellings, private libraries, business offices, and many governmental bureaus not open to the public. Such collections often yield priceless geographical information. Since there are virtually no published guides to the materials in closed collections, it must suffice here merely to remind the reader of their importance. The geographical student should never forget the possibility that, after he has exhausted the resources of open collections, the data that he most needs in connection with a particular subject may lie hidden in some closed collection of which he may have no knowledge. To gain this knowledge personal contacts are necessary. If you can once find out where materials exist in closed collections you need never hesitate to ask for permission to

---

[17] Classic works on this subject are E. Bernheim, *Lehrbuch der historischen Methode,* 5th and 6th ed. (Leipzig, 1908), and C. V. Langlois and C. Seignobos, *Introduction aux études historiques,* 4th ed. (Paris, 1909; English translation by G. G. Berry, *Introduction to the Study of History,* London, 1898; reprinted 1912). For additional references see L. J. Paetow, *A Guide to the Study of Medieval History,* rev. ed. (New York, 1931), pp. 16–17.

[18] J. K. Wright, "Map Makers Are Human: Comments on the Subjective in Maps," *Geog. Rev.,* 32 (1942), 527–544.

consult them, and frequently such permission is forthcoming, since closed collections are not necessarily secret.

Open collections are of three main types: libraries, public archives, and museums (also botanical and zoological gardens). Certain publications describe such collections in broad terms. It is not always easy, however, to ascertain whether specific items exist in particular collections without writing their librarians or curators.

Libraries [19] are either "general" or "special." A general library is one that aims to cover all fields of knowledge and all parts of the earth, or one of which the potential scope, at least, is unrestricted.[20] Special libraries are restricted to particular regions or other subjects.[21] General libraries range in size from the immense Library of Congress with its millions of books down to the village public library with a few hundred. Only the very largest have sufficiently representative collections to be altogether adequate for detailed scholarly research in every branch of geography, though smaller general libraries frequently possess special collections [22] sufficient for researches concerning particular geographical subjects.

[19] Mudge-Winchell [1], a 420–425, c 81, d 76–79.

[20] The American Library Association Board on Resources in American Libraries has recently investigated the leading libraries of the United States with a view to determining which contain outstanding collections. A list of 76 subjects was drawn up, and about 500 authorities were asked to state where, in their opinions, the best library collections in these fields are held (R. B. Downs, "Leading American Library Collections," *Library Quarterly,* 12 (1942), 457–473). As rated by these authorities the leading ten libraries of the country, in terms of the number of fields represented by outstanding collections, were, in descending order: Harvard, Library of Congress, University of California, Columbia, University of Michigan, New York Public Library, University of Pennsylvania, University of Chicago, Yale, Cornell.

[21] See *Special Libraries Directory of the United States and Canada* (New York, Special Libraries Assn., 1935), 253 pp. In the main list, arranged by localities, data are given for each library concerning its staff, size of collections, the subjects represented, special collections, etc. Subject index. Does not deal with special collections in general libraries.

[22] E. C. Richardson, *An Index Directory to Special Collections in North American Libraries . . .* (Yardley, Penn., F. S. Cook and Son, 1927), 168 pp. Contains two alphabetical indexes, one by localities, the other by subjects. For most of the collections the approximate number of volumes, but no other information, is given.

For 76 subjects classified under the following main headings, Downs ("Leading American Library Collections") lists American libraries in which there are outstanding

While a special library may not contain a larger or better collection of works on its field than may be found in the largest libraries, a special library is usually more conveniently organized for studies in that field. It is likely to be catalogued in greater detail and according to a system of classification adapted to the field. Its books and other documents may be consulted more rapidly because they are concentrated in a small space instead of scattered through an immense building. On the other hand, special libraries do not usually possess many works marginal or auxiliary to the specific study that the student is likely to pursue in them and to which he may well wish to refer.

The special libraries and special collections within general libraries that are of particular value to geographers are, naturally, those devoted to geography [37, 38] and allied subjects. Such collections are often arranged or catalogued by places and regions as well as by other subjects. Regional collections may consist largely or exclusively of materials on the history, art, literature, manners and customs, or natural history of a region and be arranged or catalogued with little or no thought regarding the geography of the region. In a truly geographical library, on the other hand, there is always a nucleus of strictly geographical documents, and in addition, a greater or lesser quantity of other material of potential value in geographical studies. While geographical libraries may contain unique and rare regional and geographical documents, copies of many of their current books are available in other libraries.

---

collections: literature and languages, history (United States, Latin American, medieval, English, French, German, Italian, Spanish, Near East, Middle East, Chinese, Japanese, First World War), fine arts, philosophy and religion, social sciences, economics, science (includes among other subjects: geology, geography, maps, botany, zoology, anthropology and ethnology). Except under history the subject headings are topical rather than regional (see p. 30).

Constance M. Winchell, *Locating Books for Interlibrary Loan: With a Bibliography of Printed Aids Which Show the Location of Books in American Libraries* (New York, 1930). The bibliography (pp. 49–170) contains a brief section on geography and also under the headings "Americana" and "History" references to printed library catalogues and union lists pertaining to specific regions and states in the United States and to certain foreign countries.

For various groups of libraries union catalogues [23] and union lists [24] have been compiled which indicate the collections in which copies of each work listed are kept, and there is a comprehensive *Union List of Serials* [25] covering the holdings of a large selection of popular, scientific, and technical serials in most of the important general and special libraries of the United States.

The entire catalogues of a few libraries have been published [26] and many libraries publish lists of current accessions. All students of geography should know that it is not necessary actually to visit the largest libraries in the world—the Library of Congress, the Harvard University Library, the Bibliothèque Nationale, and the British Museum—in order to consult their catalogues. In whole or in part these are available for reference (in the form of cards or of books) in many of the large libraries of this country.

Public archives—collections of manuscripts of permanent historical value—are of particular interest to students of the historical aspects of geography. The National Archives of the United States in Washington, devoted to manuscript and other materials from the files of the various federal departments, contains many maps and geographical reports, and a professional geographer is in charge of its map department.

[23] R. B. Downs, ed., *Union Catalogues in the United States* (Chicago, Amer. Library Assn., 1942). Contains a "Directory of Union Catalogs in the United States" (pp. 351–391) and a selected bibliography (pp. 395–402).
[24] See Downs, *Union Catalogues*, pp. 6–7.
[25] Winifred Gregory, ed., *Union List of Serials in Libraries of the United States and Canada*, 2d ed. (New York, 1943), 3,065 pp. Covers period through 1940 (*Supplement, 1941–1943*, published in 1945, 1123 pp.). Indispensable for verifying titles of periodicals of all types both American and foreign, for determining periods covered by their issues, and especially for ascertaining where files may be found. For each periodical the volumes held in certain important American libraries are indicated. Contains "A Bibliography of Union Lists of Serials" by D. C. Haskell and Karl Brown (pp. 3053–3065).
[26] Mudge-Winchell [1], a 423–425, b 57, c 81, d 79. For a convenient catalogue of a general library published in book form see C. T. Hagberg Wright and C. J. Purnell, *Catalogue of the London Library*, 2 vols. (London, 1913, 1914); Supplements, 2 vols. (1920, 1928); *Subject Index of the London Library* . . . , 2 vols. (1909, 1923). The last is alphabetical as regards the main subject headings, with copious cross references. For published catalogues of geographical libraries see [8, 9, 16, 17], below.

Although geographers are not in the habit of making much use of museums [27] and botanical and zoological gardens, an immense amount of geographical information has nevertheless been gleaned from their collections—information with regard to the distribution and interrelationships of the physical features of the earth's surface, of plant and animal species, and of human cultures. The utility of a museum for purposes of geographical study depends largely upon the manner in which its exhibits are arranged. In many modern museums there has been an increasing emphasis upon regional arrangement as distinguished from arrangement by orders, species, and genera, or by historical periods, or other nongeographical criteria. Regional museums—museums devoted to the natural history or culture of particular regions—have also been maintained in many places, although they are not ordinarily directed by geographers and their exhibits are sometimes heterogeneous. We have no strictly *geographical* museum in this country,[28] though there are a few small geographical exhibits in certain of our university departments of geography. Before the war there were two distinctively geographical museums in Europe—at Leipzig [39] and Leningrad [40]—but whether they have survived is not known to the writer. Geographical museums and exhibits contain such geographical documents as flat and relief maps, globes, and models. They also contain photographs, landscape paintings, dioramas, and specimens of many kinds. Their effectiveness depends on the selection, arrangement, and labeling of the objects, most of which in themselves do not differ intrinsically from objects found in nongeographical exhibits.

The use of bibliographical aids.—The student making up a list of references to publications to be consulted in connection with a geographical study may adopt either a hit-or-miss method or the more systematic procedure of going through bibliographical aids. The hit-

[27] Mudge-Winchell [1], a 37–38.
[28] The St. Paul Museum, St. Paul, Minn., has recently developed some striking exhibits of globes, maps on spherical surfaces, and map projections (see L. H. Powell, "New Uses for Globes and Spherical Maps," *Geog. Rev.* [476], 35 (1945), 49–58).

or-miss method is to compile references solely from the text or footnotes of books and articles bearing on the subject under investigation. One work leads to another, and in the course of time a considerable quantity of references may be amassed. The method obviates disagreeable searching through bibliographies and pawing over card catalogues and for some subjects is sufficient, but it has disadvantages. The bibliographer and the writer approach the literature of a particular subject from different angles. The writer is interested primarily in the subject itself. Most of his references in footnotes are intended to substantiate assertions in his text. He is likely to regard fussing with bibliographical details as a nuisance and for this reason to make mistakes and omissions. The bibliographer, on the other hand, is interested in the literature of the subject for its own sake. His bibliographical instinct is like that of the collector of books or works of art, an instinct that impels him to strive for completeness and to regard with almost loving care each reference in his collection. The bibliographer's enthusiasm for books as books, articles as articles, maps as maps, may, however, exceed his critical knowledge of their subjects and lead him to list worthless or unimportant publications. Even so, his bibliography will usually be of greater service than are most casual footnote references in regard to the correct spelling of names, titles, and places, and in its consistency and accuracy as to dates, number of pages, and the like.

Casual references may, of course, disclose pertinent materials not listed in the available bibliographies and library catalogues, and no serious student can afford to ignore the casual references in books and articles on his subject. It is only when he relies on them alone that his method is hit-or-miss.

Geographers are sometimes asked to recommend a "bibliography of geography" covering the entire subject. No such work exists. Indeed, there is no single comprehensive up-to-date bibliography giving references even to the writings of professional geographers for all regions and all branches of geography, and a complete bibliography of all publications of potential value to the geographer would have to embrace a large proportion of everything in print.

Geographical studies, like studies in history, require the use of very diverse bibliographical aids. In this respect they differ from studies in many other fields. The research worker in classical archeology, meteorology, or ornithology, for example, may consult, if not a single bibliography for the whole of his field, at least a limited number of aids from which he may obtain virtually all the references he is likely to need. The geographer can seldom rely on geographical bibliographies alone. While he need not try to acquire a profound knowledge of the whole realm of bibliography, he should know the principal landmarks in this realm, in order that he may pick up a trail and follow it without having to depend too largely on the guidance of librarians. The librarians of large libraries may render great help, but they are unlikely to be familiar with the specific applications of bibliographical material to geographical purposes, and, unfortunately, geographical studies cannot always be pursued in geographical libraries.

Bibliographical aids vary as to the quantity and quality of the information that they furnish. For almost any broad subject, aids that aim to be complete usually provide less detailed information about the individual items than do aids that are more critically selective. The information concerning specific items given in comprehensive aids and those intended mainly for the use of librarians and bibliographers is likely to include more details about the external and physical characteristics of a work than about its contents. Special aids for the scholar and scientist are better guides to quality. For the use of scientists bibliographies that include the titles of books only are normally of much less value than those that also contain references to periodical articles, since many of the most significant results of scientific research never find their way into books. Other things being equal, a bibliographical aid that furnishes information about separate maps and maps in books and periodicals is of more use to the geographer than one that fails to do so. Book reviews and bibliographies in which the references are selected and annotated are of value when prepared by persons familiar with the subjects with which they deal. But book reviewers have been known to be ignorant or prejudiced, and the maker of even a critical bibliography cannot be expected to have a thorough mastery of the entire subject that he attempts to cover. A bibliographical aid, like any other document, must be used with discrimination.

Classification and arrangement of the material in bibliographical aids.—The utility of a bibliographical aid depends hardly less on the manner in which its materials are classified and arranged than on the nature of the materials themselves.[29] By *classification* is meant the grouping of the materials according to classes; by *arrangement,* the order in which the several classes and the individual items within each class are presented. The arrangement may or may not be related to the classification. Obviously there is no such relationship where the arrangement is alphabetical. In most bibliographies the classification of the titles is either by author or by subject; occasionally it is based on other criteria, such as date, publisher, place or form of publication. Frequently different principles of classification are followed in different parts of the same bibliography.

For purposes of geographical research it is important to distinguish between three types of subject classification.

First, there are *place* or locational classifications in which the materials are grouped according to the names of specific geographical places, regions, or features—countries, states, mountains, oceans,

[29] On geographical classification see S. W. Boggs, "Library Classification and Cataloguing of Geographic Material," *Annals Assn. Amer. Geogs.* [479], 27 (1937), 49–93 (noted in *Geog. Rev.* [476], 27 (1937), 688–689; see also [20]); F. F. Outes, *La determinación de las fuentes de la geografía nacional: Agrupación sistemática de la bibliografía geográfica y regesta cartográfica de la República,* constituting Facultad de Filosofía y Letras de la Universidad Nacional de Buenos Aires, *Publicaciones de la Sección de Geografía* . . . No. 3 (Buenos Aires, 1921); also two papers on the same subject by the same author in *Anales de la Sociedad Científica Argentina,* 88 (1919), 173–209. See also [107].

rivers, capes, cities. A regional (or areal) classification is one variety of place classification. Place classifications are commonly called "geographical," but the former term is preferred in order to distinguish them from geographical classifications of the third type, described below.

Second, there are *topical* classifications, in which the materials are grouped according to nonlocational criteria. Any fact or concept that is not related to a specific place may, somewhat arbitrarily, be designated as a "topic." Geographical topics may represent features of the earth's surface when considered generically rather than specifically. Thus, in a topical classification some of the material might be grouped under such headings as "Mountains," "Rivers," and so on, whereas in a locational classification such headings as "The Alps," "The Amazon," would appear. Classifications by bibliographical subjects, which will be discussed in the next section, form one variety of topical classification.

Third, there are *geographical* classifications. These are designed primarily to meet the needs of geographers or others engaged in geographical research. They are not distinct from or coördinate with classifications of the first two types but may be either by places or by topics, or else, as in the present volume, in part by places and in part topical. The place subjects in geographical classifications do not differ essentially from those of other place classifications. The topics usually comprise the various topical aspects or branches of geography (physical, human, economic, and the like) and other topics in which geographers are likely to be interested.

Where the classification of the materials in a bibliographical aid is either topical or by places, the arrangement may bear some definite, logical relationship to the scheme of classification. Thus, if the classification is by places, the references might be entered by major headings for continents with subheadings for the countries in each continent, and minor subheadings for the states and provinces in each country, the order of the various headings and subheadings being determined by the relative positions of the several areas (for example, the areas could be arranged in "tiers" from north to south, with the headings within each tier from west to east). A logical arrangement of this sort may be designated a *locational* arrangement. Similarly, where the classification is topical, the arrangement may be by major topics, subtopics, and minor subtopics (for example, orders, genera, species), the sequence of the various topics and subtopics being based in so far as possible upon some relationship between them. Such logical arrangements are designated in this book as *systematic*.

On the other hand, the arrangement may be wholly unrelated to the classification. For example, a classification might be by places or topical, with the subject headings arranged alphabetically or chronologically.

Bibliographical aids are often used for looking up details, such as the title, date, publisher, and price of a work of which the author's name is known, or the author of a work of which the title is known. For such purposes classification by author and title with alphabetical arrangement is obviously the most convenient; for this reason, also, an aid in which the arrangement is logical is usually accompanied by an alphabetical author index. Again, if the student remembers that a work exists on a certain specific subject but has forgotten its author and title, a reference to the work in question can often be more easily found in an aid or index where the subjects are arranged alphabetically than in one in which they are arranged logically.

The principal disadvantage of alphabetically arranged subject classifications (including those of the "dictionary catalogues" found in most libraries) lies in the fact that they tend to scatter the references pertaining to many related subjects. In them, for example, under such subject headings as "New England" or "Transportation" will ordinarily be found references solely to works that deal with New England as a whole, or with transportation as a whole, or in the titles of which the words "New England" or "transportation" happen to appear. These, however, might comprise only a small fraction of all the references relating to New England or to transportation, for example, those classified under "Maine," "Boston," "White Mountains," "North Atlantic states," or under "roads," "railroads," "airways," and so on. In other words, the items entered under a subject heading in an alphabetically arranged aid are usually restricted to those that bear on that specific subject as a whole and do not include those that bear on parts of the subject or on larger subjects of which the subject itself is a part. This may be a considerable disadvantage for the student of geography who wishes to compile a list of references pertinent to a whole region or whole topic. Skillful cross-referencing helps overcome some of this difficulty, but cross-referencing is seldom worked out so thoroughly as to insure against the risk of overlooking references of importance.

Aids in which the material is arranged locationally or systematically are therefore nearly always of greater value for use in compiling lists of references on a given subject than are alphabetical aids. There is, however, no universal scheme of classification and arrangement suited to *all* types of research. Such comprehensive schemes as the Dewey decimal system or the Library of Congress system have many advantages for general purposes but seldom satisfy the specialist. On the other hand, classifications designed especially for one field of investigation may not work at all well in another. For example, references to publications on nutritional plants of interest to the agronomist or agricultural geographer would be widely scattered throughout a botanical bibliography classified according to taxonomic principles.

Hence, no matter how well organized and carefully indexed it may be, a bibliographical aid will never completely suit all who may consult it. No bibliographer can foresee and provide for all the possible subjects or combinations of subjects with which the users of his bibliography might conceivably be concerned. Since geography deals largely with regions and areal distributions on the earth's surface, the effectiveness of a bibliographical aid as a tool for geographical research depends largely on its organization in these respects. If its organization is such that one may easily run down references to works dealing with the specific regions or with the geographical distributions in which one is interested, a bibliography or catalogue is usually of more use to the geographer than one that obliges him to make protracted searches.

THE PRINCIPAL TYPES OF BIBLIOGRAPHICAL AID

a) *Bibliographical aids considered in terms of their bibliographical subjects.*—Every bibliographical aid to geographical research may be said to deal with both a geographical and a bibliographical subject. The geographical subject is the aspect of geography—it may include the entire field—covered by the documents about which the aid provides information. The bibliographical subject is the kind or category of documents with which it deals, as determined by criteria unrelated or only indirectly related to their subject matter. The bibliographical subject of a guide to United States government documents on climatology is United States government documents; the geographical subject is climatology. In this section bibliographical aids will be considered in terms of their bibliographical subjects. In connection with the various types mentioned, references will be given to certain universal aids and comprehensive aids of especial value to geographers, as well as to the pages in Mudge-Winchell where additional data may be found.

The criteria most frequently applied in defining the bibliographical subjects of aids to research are of six kinds: criteria of (1) representation in libraries, (2) date of issue, (3) place or language of publication, (4) producer, (5) mode of issue, and (6) form or purpose. These, of course, are not mutually exclusive. Indeed, criteria of all six types might conceivably be applied together, as, for example, in a hypothetical list of educational (6) periodicals (5) in the Library of Congress (1) published by commercial publishers (4) in the United States (3) before 1850 (2).

1) Representation in libraries.—Library catalogues, union cata-

logues, and union lists are the principal bibliographical aids for which representation in libraries is the basic criterion of selection. Such aids, when published, may often be used as bibliographies as well as for the purpose of finding out whether one or more specific libraries possess the works listed in them. The published catalogues of the world's great libraries, to which we have already referred (p. 25), are fundamental bibliographical tools, and many catalogues of special libraries are also of considerable value for research purposes [20, 21]. By and large, however, a library catalogue is likely to contain more heterogenous material and material of more uneven quality than do critically selected bibliographies covering the same subjects.

2) Date of issue.—Bibliographical aids may be either current or retrospective. Current aids, issued periodically or at frequent intervals, record contemporary publications. Retrospective aids aim to furnish complete or selected references to publications that have appeared during a specified or indefinite period in the past. In actually using bibliographical aids of either type to compile a list of references, the student will frequently find it advisable to begin with the most recent ones and work backward chronologically as far as seems necessary.

3) Place of publication or language.—In nearly every civilized country books published in that country are regularly recorded in current trade repertories that appear periodically at frequent intervals and are published in cumulative form every now and then.[30] For certain countries and also for certain languages there are also retrospective lists and bibliographies of books that have been printed in those countries or languages during specified periods. Such aids [31] may be used to advantage in order to ascertain titles, authors' names, places, dates, publishers, prices, and the like, but for purposes of research in compiling lists of references their value is more questionable,

[30] Lawrence Heyl, *Current National Bibliographies: A List of Sources of Information Concerning Current Books of All Countries,* rev. ed. (Chicago, Amer. Library Assn., 1942), mimeographed.
[31] Mudge-Winchell [1], a 379–413, b 53–56, c 75–81, d 71–75.

since they give an immense quantity of titles, including many of little or no significance.

4) Producer.—The producers of a document are its author, its publisher, and sometimes a sponsor as well. Obituary notices and biographical memoirs,[32] the various biobibliographical reference works,[33] and the author entries in card catalogues and bibliographies provide information regarding the works of individual authors. The sponsor of a publication may or may not be identical with its publisher. Many institutions, such as universities, sponsor works that they do not publish—as in the case of doctors' and masters' dissertations (which may be of considerable utility in geographical research [33] and for which lists are available).[34] Commercial publishers' catalogues, lists of the publications of congresses,[35] international organizations, scientific societies, and other learned institutions,[36] and guides to government documents [37] are bibliographical aids for which

---

[32] Short obituary notices of geographers and explorers are published in the *Geographical Review* [476] and other geographical periodicals. More extensive biographical memoirs of late members of the Association of American Geographers will be found in the *Annals of the Association of American Geographers* [479]. References to obituary notices and biographies of geographers are given in the *Bibliographie géographique internationale* [7], in *Current Geographical Publications* [9], and in the numbers of the *Geographisches Jahrbuch* [6] covering 1888–1903 (continued in *Geographen Kalender* [725] through 1914). See also the geographical encyclopedias of Banse [70] and Kende [69] containing short articles on a few outstanding geographers and explorers. Paul Lemosof, *Le Livre d'or de la géographie: essai de biographie géographique* (Paris, 1902), is a biographical dictionary of "explorers, geographers, [and] cartographers who have contributed to the actual state of our knowledge of the earth from the earliest times to the end of the nineteenth century" (the comments on each individual are seldom more than three or four lines in length).

[33] Mudge-Winchell [1], a 281–314, b 39–45, c 57–65, d 55–62.

[34] *A List of American Doctoral Dissertations Printed in* [1912–1938] (Library of Congress) (Washington, annual).

Doctoral Dissertations Accepted by American Universities, 1933– (New York, H. W. Wilson, 1934– [annual]).

Mudge-Winchell [1], a 29–33, b 12–13, c 14–15, d 12–13.

[35] See note 11.

[36] Mudge-Winchell [1], a 34—37, c 13, d 11–12. Short reports on international geographical congresses and other congresses of geographical interest are published in the *Geographical Review* [476].

[37] J. B. Childs, *Government Document Bibliography in the United States and Elsewhere,*

the criterion of selection is the publisher. Of these, guides to government documents are by far the most valuable in geographical studies, since the material in government documents is often inadequately catalogued in libraries and is seldom fully listed in bibliographies.

5) Mode of issue.—Some documents are issued as wholly independent units; others as parts of sets, series, or serial publications. For many of these there are lists and indexes. Since most of the books and monographs that form parts of series are catalogued and entered in bibliographical aids as if they were independent items, the student of geography is unlikely to have much need of lists of the items in such series unless he wishes to check an incomplete series in his own or some other library, or to use the lists in connection with a study of the work of the institution that has sponsored or published the series listed.

In the case of serial publications the problem is different. Serial publications, which are issued at regular or more-or-less regular intervals, are of two kinds: (1) serials that systematically present in successive issues the same sort of information brought up to date, as, for

---

3d ed. (Library of Congress, Division of Documents) (Washington, 1942), 96 pp. Offers a quick means of determining what bibliographical tools are available for specific countries and for the states of the United States.

Anne M. Boyd, *United States Government Publications: Sources of Information for Libraries,* 2d ed. (New York, H. W. Wilson, 1941), 548 pp. A running discussion of the organization of the government, of the nature of its publications, catalogues and indexes, and of the publications of Congress, the several executive departments, and independent establishments, with copious bibliographical data.

Winifred Gregory, ed., *List of the Serial Publications of Foreign Governments, 1815–1931* (New York, H. W. Wilson, 1932), 720 pp. This monumental work, edited for the American Council of Learned Societies, American Library Association, and National Research Council, lists alphabetically by country the serial publications of national governments and of such lesser states as are to some extent self-governing. The holdings in 85 important American libraries are indicated.

J. K. Wilcox, ed., *Manual on the Use of State Publications* (The Committee on Public Documents of the American Library Association) (Chicago, Amer. Libr. Assn., 1940), 352 pp. This comprehensive volume contains a bibliography of bibliographies of state publications, pp. 75–91).

See also Mudge-Winchell [1], a 364–374, b 52, c 72–74, d 69–70.

example, yearbooks, almanacs,[38] and current bibliographies; and (2) periodicals and other serials (such as transactions of learned societies) which present a variety of separate items that differ from issue to issue.

For both types of serial there are reference works listing the serials by topics or place of publication and giving data with regard to their character, editors, publishers, subscription prices, and so on.[39] Likewise the individual volumes of both types are usually provided with indexes, and for many serials cumulative indexes to a run of volumes are issued from time to time.[40] In addition, for certain selected groups of periodicals, including not only popular magazines but also many scholarly and scientific journals, there are bibliographical indexes [41] in which the student may look up references to articles or book reviews,[42] alphabetically arranged by author, title, and subject. Since material in periodicals is seldom catalogued in general library catalogues, indexes of this kind, particularly the indexes to geographical periodicals,[43] may be useful indeed in geographical studies.

6) Form or purpose.—Finally, the bibliographical subject of a bibliographical aid may be restricted in scope to works of particular physical forms, such as manuscripts, printed texts, bound volumes, or

[38] Mudge-Winchell [1], a 121–129, b 21–22, c 25–26, d 24–25. For many regional yearbooks not listed in Mudge-Winchell, see the bibliographies in the *Statesman's Year-book*. There are also many yearbooks which pertain to particular organizations (e.g. governments, political parties, societies, etc.). Some of these are listed elsewhere in the present work; for others Mudge-Winchell should be consulted.

[39] Mudge-Winchell [1], a 17–23, b 10–11, c 10–11, d 9.

[40] D. C. Haskell, *A Check List of Cumulative Indexes to Individual Periodicals in the New York Public Library* (New York, New York Public Library, 1942), 370 pp.

[41] *Readers' Guide to Periodical Literature* (New York, H. W. Wilson, 1905–). Materials in the better-known general and popular periodicals of America are indexed by author and subject.

International Index to Periodicals, Devoted Chiefly to the Humanities and Science (New York, H. W. Wilson, 1907–). Materials in foreign as well as American periodicals of the more scholarly type, including practically all of the more important geographical and related periodicals, are indexed by author and subject.

See also Mudge-Winchell [1], a 6–17, b 9–10, c 9–10, d 7–9.

[42] Mudge-Winchell [1], a 13, b 10, c 10, d 8.

[43] Data concerning these are given in the references to geographical periodicals in the Bibliography, below.

pamphlets, or it may be restricted to works in which information and ideas are expressed through particular media, as in the case of texts, mathematical tables,[44] maps [92–107], or photographs [8, 9]. A bibliographical aid may deal exclusively with works of certain specified literary forms, such as nonfiction, novels [30, 32], or poems, or with works designed to serve certain specific purposes, as in the case of scientific monographs, census reports and other statistical works,[45] atlases [108–131], textbooks [22–29], guidebooks [88–91], gazetteers [65–73], indexes,[46] and bibliographies [3–5].[47] Some of these criteria of form or purpose are wholly unconnected with any geographical subject; a novel or a manuscript may deal with almost anything under the sun. Others, however, are inherently geographical as well as bibliographical; maps, gazetteers, and guidebooks necessarily deal with geographical subjects.

b) Bibliographical aids considered in terms of geographical subjects.—In terms of their geographical subjects, the aids listed in Part II are topical but not regional. With the exception of the geographical periodicals, those listed in Part III are all regional, though in many

[44] Mudge-Winchell [1]: comprehensive scientific tables, a 170; mathematical tables, a 170–172, b 27, c 35, d 32; astronomical and navigational tables, a 172–174; physical and chemical tables, a 175–176, 179, b 28; weights, measures, money, a 153–154, c 31.

[45] References to statistical works and to bibliographies dealing with them will be found under the following main headings: dictionaries (of statistics), almanacs and general yearbooks, census reports, statistical abstracts and national yearbooks, in Mudge-Winchell [1], a 120–129, b 21–22, c 25–26, d 23–25. See also Hirshberg [2], 1942, pp. 203–208; Besterman [3], 2 (1940), 342.

[46] Norma O. Ireland, *An Index to Indexes: A Subject Bibliography of Published Indexes* (Boston, F. W. Faxon, 1942), 107 pp. Covers "special indexes," indexes to sets of books, periodical indexes, cumulative indexes to individual periodicals, and government document indexes.

See also notes 40–41, above.

[47] *The Bibliographic Index: A Cumulative Bibliography of Bibliographies* (New York, H. W. Wilson, 1938–). Quarterly, with cumulative annual volumes; also general volumes covering 1937–1942 and 1943–1944. Covers brief incidental bibliographies in books and periodicals as well as separately published bibliographies. Alphabetically arranged topical and regional headings. The material under the regional headings is often subdivided by topics and vice versa.

See also Mudge-Winchell [1], a 376–377, b 53, c 74, d 70–71.

instances they are confined to particular geographical or bibliographical topics concerning the several regions.

The *topical aids* of greatest service in geographical research are (1) geographical aids to research in the various branches of geography and (2) nongeographical aids designed for use in related studies in which a large measure of attention is given either to geographical distributions, or to the nature of the regional subdivisions of the earth's surface, or to the question of human and other forms of life in relation to the natural environment. Aids of both types are listed in Part II, and their classification may be readily understood by looking over the headings in Part II itself. It often happens, however, that the student of geography may wish to carry his researches so far into the domain of a related field that the aids referred to in Part II will not suffice and he must turn to other bibliographical tools. A number of fields into which geographers are likely to have occasion to penetrate are listed below. Each of these studies is sufficiently well developed to be represented by at least one scientific or scholarly journal in the English language, and many are represented by several journals.

Fields pertaining primarily to the natural environment: science in general; natural sciences in general; geodesy; geophysics; geology; glaciology; economic geology; petroleum geology; geomorphology; soil science; meteorology; climatology; oceanography; biology; ecology; botany; plant ecology; zoology; animal ecology.

Fields pertaining primarily to man: history in general; ancient history; medieval history; modern history; agricultural history; maritime history; economic history; history by countries; history of particular regions within countries; archeology; anthropology; ethnography; economics; agricultural economics; land economics; sociology; political science; military science.

Fields pertaining to techniques of acquiring, applying, disseminating geographical knowledge: surveying; photogrammetry; statistics; navigation; education.

A regional aid to geographical research may cover a limited number of geographical or bibliographical topics or else may seek to cover a very wide range. Normally the larger and more complicated the region, the more unsatisfactory is a regional aid that aims to deal with all manner of works on all manner of subjects relating to it. Such aids are never complete and seldom critical.

A bibliographical aid may be regional in effect even though nominally it does not deal with a region, as such. The publications of city, provincial, and state governments and also of those agencies of national governments concerned mainly with interior affairs and resources (for example, the United States Land Office, United States Geological Survey) deal for the most part with particular regions, and bibliographies of such publications are essentially regional. Much the same is true of the publications of small learned institutions, local geographical societies, and the like.

It is also not easy to draw the line between geographical periodicals that are actually general in scope and those that are nominally general but actually regional in scope. Partly because of this difficulty and partly because in all general geographical periodicals somewhat more emphasis is placed on the home country than on other parts of the world, we have listed general geographical periodicals in Part III rather than in Part I, where they might more logically be placed.

The classification and arrangement of the material in Parts II and III approximate those of the Research Catalogue of the American Geographical Society though [20], modifications have been made. They do not purport to reflect an entirely logical or philosophical organization of the subject matter of geography. Geographical studies as a whole advance in an inconsistent manner along an irregular front. Pioneers are constantly opening up new realms of investigation that do not fit into preconceived categories. The subdivisions in the classification aim to reflect subjects in which geographers are working or have worked. The general rule has been adopted that, if a publication deals with a region, it is classified under the region, since regional studies are the distinctive province of geographers. However, material on the history of geography, which forms a somewhat exceptional branch of geographical research, has been included together in one place in Part II even though some of this material might have been listed in Part III under regional headings or under other topical headings in Part II.

BIBLIOGRAPHY

EXPLANATORY NOTES

Order of references under main headings.—The scheme of classification adopted for the bibliography is explained on pp. 14–16, above, and its framework is made clear in the Table of Contents.

Under each main heading in Parts II and III the entries, in so far as they are provided, are normally arranged in this order:

1) Bibliographies of books, periodical articles, and the like (often preceded by introductory matter),

2) Periodicals and series,

3) Bibliographies of maps and atlases; map indexes,

4) Atlases.

The principal references of each category are normally arranged in the reverse chronological order of their dates of publication, since the most recent works should ordinarily be consulted first. This rule has not been followed where a different order is more helpful. For example, works dealing with different historical periods are listed chronologically by the periods to which they pertain.

Dashes are used to separate minor categories of references.

References to current periodicals and series precede references to those that have been discontinued. The former are listed in chronological order according to the dates of their first numbers (or those of the first numbers of their earliest direct predecessors, as in the case of the *Geographical Review* [476]). All periodicals and series that were appearing in 1939 and are not definitely known to have been discontinued are regarded, for present purposes, as current. Discontinued periodicals (that is, those either known to have been discontinued or for which there is no record at the American Geographical Society of numbers published since 1938) are listed in inverse chronological order according to the dates of the latest (or latest-known) numbers.

Cross references to the serial numbers of works cited elsewhere in the Bibliography are indicated by figures in square brackets. Cross references are *not* provided from the sections dealing with smaller subjects to those dealing with the larger subjects of which the smaller subjects form parts—or vice versa. For example, there are no cross references from the section on Venezuela to works listed under "South America" or "The Caribbean Area" that deal with Venezuela among other neighboring countries, nor are there cross references from "South America" to "Venezuela." Hence, in order to make the best use of the Bibliography the reader should not only look up the cross references given in the section covering a topic or area in which he is immediately interested but should also consult the Table of Contents for guidance to other sections that might be pertinent. He should also remember that the most complete or the most recent reference work covering a given subject is not necessarily listed under the subheading for that specific subject, as it may frequently be a work that embraces a more comprehensive subject.

Main references.—These are self-explanatory except as follows: Wherever the author of a work is known to be a woman, her first name, if known, is given in full. Initials only are given for the first and middle names of men having three or more names.

Where a work has a title in two languages, both titles are usually given in italics. Translations of certain Russian and other titles, where supplied by the compilers of the present volume, are given in roman type in square brackets.

Periodical titles are given in their most recent forms. Minor divergencies from these forms are in some cases indicated in parentheses within the main title. Other changes in titles are explained in the annotations.

The sponsors of works (see above, p. 35) are named in parentheses after the titles. The publisher of works sponsored by agencies of the United States Government is usually the Government Printing Office.

The number of pages indicated is either the number carrying arabic numerals only or, in most instances, the approximate total of all pages carrying arabic and roman numerals.

Where the date of termination of a publication is not definitely known, the most recent date on record at the American Geographical Society is given in square brackets.

Annotations. The annotations on the bibliographies are descriptive. They present information about (1) subject matter, classification, and arrangement, (2) approximate number of entries (only where the entries are numbered serially), (3) annotations, (4) indexes. They do not seek to appraise the bibliographies in terms of accuracy, completeness, selectivity, and other scholarly qualities. Critical, as distinguished from purely descriptive, comments have not been attempted. Such comments can be written fairly only by one who is thoroughly familiar with the subject that a bibliography covers—indeed only by one who knows the subject even better than does the compiler of the bibliography, and no one could be such an authority on all of the many subjects covered in this book.

The present compilers have resisted the temptation to characterize bibliographies as "important," "useful," "valuable," and so on. These terms are not particularly helpful, since no bibliographies not deemed of some potential value and importance are listed. "Importance" or "utility" are relative matters.

Among the facts that the geographer is likely to wish to ascertain about a bibliography are whether or not it contains references to periodical articles and maps as well to books, whether or not it contains references to other bibliographies, and whether or not its scope, classification, and arrangement are such as to facilitate its use in geographical studies—particularly in the assembling of references pertaining to particular places and regions. The annotations are designed to shed light on these matters.

Since the great majority of the bibliographies listed were compiled for use as tools of research in geography and related fields, most of them include references to articles in periodicals as well as to books, and this may be assumed, except where the title or annotation indicates specifically that books or periodical articles only are included.

Either a part or the whole of each annotation consists of comment in which certain terms and abbreviations are used as defined below. In this connection, the remarks on classification and arrangement in the Introduction should also be consulted (pp. 29–33, above).

alph.: alphabetical. The abbreviations "alph.," "chron.," and "system." are descriptive of the arrangement (see p. 29) of bibliographies and indexes or parts thereof. Such an abbreviation is placed either (1) in parentheses immediately following the item to which it applies, or (2) where it applies generally to several items, at the beginning (for example, "alph. by topics subcl. by countries" means that the topical headings as well as the subheadings for countries under each topical heading, are arranged in alphabetical order). "alph." is ordinarily omitted where the reference is to an alphabetical index or alphabetical author classification, since the arrangement of indexes and author classifications is normally alphabetical. Where no indication is given of the arrangement of a bibliography (or a part thereof), the arrangement is not deemed of sufficient practical importance to require comment.

ann.: annotated, annotation (where "ann." is omitted annotations are either wholly lacking or scanty).

art.: article, paper, etc., in a periodical or other serial.

auth.: author.

bibl.: bibliography, bibliographical.

by: (see "cl.").

ca.: circa, approximately.

chron.: chronological, chronologically (see "alph.").

cl.: classified, classification. The nature of a system of classification may be indicated in either of two ways: e.g. "topical cl." or "by topics"; "regional cl." or "by regions"; "place cl." or "by places."

entry: a reference in a bibliography to a particular work.

genl. index: an index in which there are captions for authors, places, and topics (see "place" and "topic").

geog.: geography, geographical (see "place").

geog. names: names of places (see "place").

geog. index: an index pertaining to the subject of geography as such, as distinguished from a place index or index to geog. names (see "place").

hdg.: heading.

incl.: includes, including, among which there will be found.

index: where an annotation contains no reference to an index, it may be assumed that an index is lacking (see also "alph." and "genl. index").

natural hist.: natural history. Implies that several subjects such as geology, botany, zoology, and the like are covered.

per.: period (in chronological sense).

period. art.: periodical article.

place: any specific locality, small or large, on the earth's surface, whether in the form of an area or region (q.v.), line (e.g. river, road), or point (e.g. cape, astronomically determined position). In bibliographies and other works the term "geographical" is often used synonymously with place (e.g. geographical classification, geographical index). To avoid confusion with "geographical" in the sense of "pertaining to geography as a subject," the abbreviation "geog." is used in the annotations only where the subject of geography as such is meant, and also in the expression "index to geog. names."

publ.: publication.

ref.: reference.

regional: pertaining to a specific region or group of regions (see "place"). A regional bibliography is one that deals with a particular region; its classification may or may not be regional. For the purposes of this book the term "regional" is applied to tracts of land or water or both, small or large, and not exclusively to regions in the more restricted sense frequently used by geographers and others.

subcl.: subclassified, subclassification.

subhdg.: subheading.

subj.: subject. A matter dealt with in a document. A subject may be either a "place" or a "topic" (q.v.). In the annotations, however, "subj." is used exclusively with reference to classifications in which *both* places and topics are represented.

system.: systematic, systematically. Refers to a logical, as distinguished from an alphabetical or chronological, arrangement of topics (see "alph.," "topic"). Usually omitted, since topical classifications are normally arranged systematically.

topic: any subject other than a specific place (e.g. history, bibliography, mountains). A topical bibliography is one that deals with a particular topic; it may or may not be classified topically. A topical classification is, strictly, a subject classification in which none of the subjects represented are places. In the annotations, however, subject classifications in which there are also a few place subjects are sometimes designated as "topical."

I. GENERAL AIDS

BIBLIOGRAPHIES OF GENERAL REFERENCE WORKS AND BIBLIOGRAPHIES

Mudge-Winchell. (a) Mudge, Isadore G. *Guide to Reference Books,* 6th ed. Chicago, Amer. Library Assn., 1936. 514 pp. (b) *Idem. Reference Books of 1935–1937* . . . , 1939. 69 pp. (c) Constance M. Winchell. *Reference Books of 1938–1940* . . . , 1941. 106 pp. (d) *Idem. Reference Books of 1941–1943* . . . , 1944. 115 pp.
ence Books of 1941–1943 . . . , 1944. 115 pp.   [1]
"A reference manual for the library assistant, research worker, or other user of library resources who needs a finger post to point out the reference tools available for some particular investigation." (b), (c), and (d) are "informal supplements" to (a). See also above, pp. 16–17.

Hirshberg, H. S. *Subject Guide to Reference Books.* Chicago, Amer. Library Assn., 1942. 275 pp.   [2]
"Alphabetic subject guide to the books needed in libraries for the answering of reference questions frequently asked. . . . Though designed primarily for the librarian, the book should also be helpful to the graduate student and others who like to find their own way among books." Deals for the most part with works in English. All of the main hdgs. (as well as most of the subhdgs.) topical; for a few topics (e.g. statistics) subhdgs. by countries; auth. index.

Besterman, Theodore. *A World Bibliography of Bibliographies.* London, published by the author. Vol. 1, 1939, 611 pp. Vol. 2, 1940, 641 pp.   [3]
Lists separately published bibls. and such bibls. published in other books as form "the bulk or at least the principal purpose" of such books. Aims to be impartially international. The approximate number of entries in each work referred to is indicated, but there are no critical anns. Alph. by subjs. (places and topics); under each main topical hdg. a further topical subcl. (e.g.: Geography, (1) Bibliographies and Periodicals, (2) History, (3) General); auth. index. "In the case of bibliographies which refer to both a subject [topic] and a country, preference is given wherever possible to the subject, which is then subdivided by countries." Since there are no cross refs. from countries to the topics under which other bibls. for the same countries are sometimes listed or to lesser places within the countries, and no entries by countries in the index, one can-

not readily compile from this work lists of all of the included bibls. pertaining to particular regions.

GEOGRAPHICAL BIBLIOGRAPHIES

Bibliographies of Geographical Bibliographies

Stein, Henri. *Manuel de bibliographie générale.* Paris, Picard, 1897. 915 pp.
[4]
The part on geog. (pp. 325–400) is cl. by regions and by topics, incl. special catalogues, geodesy, climatology, physical geog., oceanography, history of geog., voyages, cartography, ethnography. The regional entries are in alph. order by countries, each of the larger countries being subdivided into sections devoted to genl. works, and to provinces or other areal divisions. Ann.; subj. index; no auth. index.

Jackson, James. *Liste provisoire de bibliographies géographiques spéciales.* (Société de Géographie.) Paris, Delgrave, 1881, 133 pp. [5] By continents with a section on genl. and miscellaneous works; includes refs. to incidental bibl. lists in published bibls.; 533 entries; auth. index.

General Geographical Bibliographies and Library Catalogues

CURRENT BIBLIOGRAPHIES

In normal times geographical works shortly after they appear are currently recorded in bibliographical guides of two types: (1) comprehensive current bibliographical repertories of geographical literature, usually appearing annually or oftener, and (2) bibliographical notes, book reviews, etc., in the periodicals and yearbooks devoted to geography and allied subjects.

The repertories [6–11] may be regarded as cornerstones of geographical research. They include innumerable references that cannot be found in the retrospective bibliographies. By examining the pertinent parts of them, particularly [6], [7], and [9], the student can often, though not always, gather all the references that he could possibly need. Unfortunately, with one exception [9], publication of the great repertories has been abandoned or postponed on account of the war.

Periodicals bring to the student's attention works that have appeared subsequently to the latest issues of the annual repertories and often give more information about the contents of works than is possible in the brief entries in the latter. Periodicals, however, cannot cover the whole field of geography as comprehensively or as systematically as do the repertories. For certain comprehensive current geographical bibliographies formerly published in geographical periodicals see [476], [625], [691].

Geographisches Jahrbuch

Ed. 1880–1929 by Hermann Wagner (see [59]); 1930–1939 by Ludwig Mecking. Gotha, Justus Perthes. 1866–[1943]. Normally annual (1915–1918 and 1919–1923), each year covered by 1 vol.; 2 vols. annually for certain recent years.   [6]
Each vol. contains a series of reports on the status and progress of study in various fields of geog., with abundant refs. to the literature. The same subjs. are not discussed in every vol., but only from time to time, some at intervals of approximately two years, some of three, and others at longer periods. Refs. to the most recent reports are given under the appropriate hdgs. in the present work and also on pp. 273–274, below.

A table of contents of recent back numbers is provided in each vol., enabling the user to find the latest preceding section devoted to any subj. Genl. system. indexes for the series will be found in Vols. 40, 1924–1925, and 52, 1937.

In addition to the reports which have appeared at more or less regular intervals and of which a key is given below, the first 3 vols. contain sections devoted to geog. chronology, geog. statistics (of population and areas), and tables of use to the geographer. All these are indexed (system.) in Vol. 10, 1884, pp. 683–692.

The following key is included here for two reasons: (1) in order to impress upon the reader's mind some conception of the extensive scope and great utility of the *Geographisches Jahrbuch;* (2) to enable him conveniently to lay his hands upon the particular vols. containing material pertinent to the subject in which he may be interested. Having once found the vol. by means of this key, further data in regard to titles, dates covered by each report, and pages upon which the report will be found may readily be determined from the admirably arranged tables of contents (at the beginning of the vol.). Figures in parentheses represent the date of the vol. indicated. The reports referred to usually deal with a period closing a year or more before the date of publication.

In the key topical hdgs. are given in English, the German terms being supplied only where there is any possibility of misunderstanding.

MATHEMATICAL GEOGRAPHY

1. Measurement of terrestrial degree: Vols. 1–10 (1866–84), 13 (1889), 15 (1891, 16 (1893), 18 (1895), 20 (1897), 23 (1900), 25 (1902), 28 (1905), 30 (1907).
2. Methods of geographical surveying: Vols. 22 (1899), 25 (1902).
3. Methods of topographical surveying: Vol. 38 (1915–18).
4. Determination of positions by geographic, geodetic, nautical, and aeronautical means: Vol. 28 (1905); the same, except by geodetic means, 36 (1913).
5. Cartography: Vols. 51–52 (1936–37), 57 (1942); map projections: 9 (1882), 10 (1884), 12 (1888), 14 (1890–91), 17 (1894), 19 (1896), 20 (1897); map projections, map drafting, map measurements: 24 (1901), 26 (1903), 29 (1906), 33 (1910). For published maps see also Nos. 72–74.

PHYSICAL GEOGRAPHY

6. Geophysics and geomorphology (*Geophysik,* the earth as a whole and the earth's surface): Vols. 8 (1880), 9 (1882), 10 (1884), 11 (1887), 13 (1889), 15 (1891), 16 (1893). See also Nos. 7 and 9.
7. Geophysics (*Physik und Mechanik des Erdkörpers*): Vols. 18 (1895), 20 (1897), 23 (1900), 25 (1902), 28 (1905), 30 (1907), 36 (1913), 38 (1915–18), 39 (1919–23), 42 (1927), 52 (1937); terrestrial magnetism: 13 (1889), 15 (1891), 17 (1894), 20 (1897), 23 (1900), 28 (1905), 36 (1913), 40 (1924–25), 44 (1929). See also No. 6.
8. Regional geology (*Neuere Erfahrungen über die geographische Verbreitung geognostischer Formationen*): Vol. 8 (1880); (*Neuere Erfahrungen über den geognostischen Aufbau der Erdoberfläche*): 9 (1882), 11 (1887), 13 (1889), 15 (1891), 16 (1893), 17 (1895), 20 (1897), 22 (1899), 23 (1900), 25 (1902), 27 (1904), 31 (1908), 33 (1910), 35 (1912), 37 (1914).
9. Geomorphology: Vols. 52–53 (1937–38,1); (*Dynamik der festen Erdrinde*): 33 (1910), 35 (1912), 37 (1914), 39 (1919–23), 42 (1927); (*Geophysik der Erdrinde*): 18 (1895), 20 (1897), 23 (1900), 30 (1907); influence of weathering and erosion on land forms: 35 (1912), 37 (1914). See also No. 6.
10. Hydrography of the land (*Gewässerkunde des festen Landes*): Vols. 30 (1907), 45 (1930), 54 (1939,1).
11. Oceanography: Vols. 7 (1878), 9 (1882), 10 (1884), 11 (1887), 13 (1889), 15 (1891), 16 (1893), 18 (1895), 20 (1897), 22 (1899), 24 (1901), 26 (1903), 33 (1910), 38 (1915–18), 40 (1924–25), 43 (1928), 48 (1933), 54 (1939, Pt. 1).
12. Meteorology and climatology: Vols. 4 (1872), 5 (1874), 6 (1876), 7 (1878), 8 (1880), 9 (1882), 10 (1884), 11 (1887), 13 (1889), 15

(1891), 17 (1894), 21 (1898), 24 (1901), 26 (1903), 29 (1906), 33 (1910), 36 (1913), 39 (1919–23), 41 (1926), 42 (1927), 44 (1929).

13. Plant geography: Vols. 1 (1866), 2 (1868), 3 (1870), 4 (1872), 5 (1874), 6 (1876), 7 (1878), 8 (1880), 9 (1882), 10 (1884), 11 (1887), 13 (1889), 15 (1891), 16 (1893), 19 (1896), 21 (1898), 24 (1901), 28 (1905), 33 (1910), 36 (1913), 38 (1915–18), 40 (1924–25), 42 (1927), 51 (1936).

14. Soils: Vol. 54 (1939,1).

15. Animal geography: Vols. 1 (1866), 2 (1868), 3 (1870), 4 (1872), 5 (1874), 6 (1876), 7 (1878), 8 (1880), 9 (1882), 10 (1884), 11 (1887), 13 (1889), 22 (1899), 24 (1901), 26 (1903), 31 (1908), 45 (1930), 53 (1938, Pt. 2).

HUMAN GEOGRAPHY

16. Ethnography and anthropology: Vols. 2 (1868), 3 (1870), 4 (1872), 5 (1874), 6 (1876), 7 (1878), 8 (1880), 9 (1882), 10 (1884), 11 (1887), 13 (1889), 15 (1891), 17 (1894), 19 (1896), 21 (1898), 24 (1901), 28 (1905), 31 (1908), 34 (1911), 47 (1932), 48 (1933).

17. Anthropogeography: Vols. 26 (1903), 31 (1908), 32 (1909).

18. Political geography and geopolitics: Vol. 49 (1934).

19. Economic and commercial geography: Vols. 50–51 (1935–36).

19a. Geography of Settlements and population: Vol. 55 (1940).

HISTORICAL GEOGRAPHY (INCLUDING HISTORY OF GEOGRAPHY)

20. Historical geography in general, including history of geography: Vols. 3 (1870), 4 (1872); history of geography, Vol. 55 (1940).

21. The ancient Greek world, including history of geography: Vols. 12 (1888), 14 (1890–91); southeastern Europe: 43 (1928).

22. The ancient world (especially Africa, Asia, Asia Minor), including history of geography: Vols. 19 (1896), 22 (1899), 28 (1905), 50 (1935).

23. The ancient Orient: Vols. 34 (1911), 43 (1928), 47 (1932); northeast Africa: 53 (1938, Pt. 1).

24. The Roman West: Vols. 34 (1911), 39 (1919–23).

25. Topography of the city of Rome: Vol. 34 (1911).

26. History of geography from the Middle Ages on: Vols. 18 (1895), 20 (1897), 23 (1900), 26 (1903), 30 (1907), 41 (1926).

27. History of cartography: Vol. 17 (1894).

EDUCATIONAL GEOGRAPHY AND METHODOLOGY

28. Methodology: Vols. 7 (1878), 8 (1880), 9 (1882), 10 (1884), 12 (1888), 14 (1890–91).

GEOGRAPHICAL CHAIRS IN UNIVERSITIES, etc.

29. Geographical chairs in universities, etc., of Europe: Vols. 8 (1880), 9 (1882), 10 (1884), 12 (1888), 14 (1890–91), 19 (1896); geographical chairs in universities, etc., of Europe and of other parts of the world: 24 (1901), 28 (1905), 32 (1909).

GEOGRAPHICAL SOCIETIES AND PUBLICATIONS

30. Geographical societies and publications: Vols. 1 (1866), 2 (1868), 3 (1870), 4 (1872), 5 (1874), 6 (1876), 7 (1878), 8 (1880), 9 (1882), 10 (1884), 12 (1888), 14 (1890–91), 19 (1896), 24 (1901), 32 (1909).

GEOGRAPHICAL NAMES

31. Geographical names: Vols. 9 (1882), 10 (1884), 12 (1888), 14 (1890–91), 16 (1893), 18 (1895), 27 (1904), 29 (1906), 34 (1911).

GEOGRAPHICAL NECROLOGY

32. Geographical necrology: Vols. 12 (1888), 14 (1890–91), 16 (1893), 19 (1896), 20 (1897), 22 (1899), 24 (1901), 26 (1903).

SCIENTIFIC JOURNEYS AND EXPLORING EXPEDITIONS

33. Scientific journeys and exploring expeditions: Vols. 1–6 (1866–76). For later period, see below, under Regional Geography.

REGIONAL GEOGRAPHY

34. North America: Vols. 12 (1888), 14 (1890–91), 16 (1893), 18 (1895), 20 (1897), 22 (1899), 25 (1902), 27 (1904), 32 (1909), 37 (1914), 45–46 (1930–31), 58 (1943); Canada, 56 (1941).

35. Latin America: Vols. 12 (1888), 14 (1890–91), 16 (1893), 18 (1895), 20 (1897), 22 (1899), 25 (1902), 27 (1904), 30 (1907), 36 (1913), 41 (1926); South America: 54–56 (1939–1941).

36. British Isles: Vols. 17 (1894), 19 (1896), 26 (1903), 29 (1906), 32 (1909), 35 (1912), 41 (1926), 50 (1935).

37. Netherlands and Belgium: Vols. 17 (1894), 19 (1896). See also Nos. 38 and 39.

38. Netherlands: Vols. 21 (1898), 23 (1900), 26 (1903), 32 (1909), 35 (1912), 38 (1915–18), 44 (1929), 54 (1939, Pt. 2). See also No. 37.

39. Belgium: Vols. 21 (1898), 26 (1903), 29 (1906), 32 (1909), 35 (1912), 38 (1915–18), 44 (1929). See also No. 37.

40. France: Vols. 17 (1894), 19 (1896), 21 (1898), 23 (1900), 26 (1903), 29 (1906), 32 (1909), 35 (1912), 38 (1915–18), 43 (1928), 57 (1942).

41. Southern Europe: Vols. 17 (1894), 19 (1896), 21 (1898), 23 (1900), 26 (1903), 29 (1906). See also Nos. 42, 43, 56–59.

42. Italy: Vols. 32 (1909), 35 (1912), 41 (1926), 46 (1931), 56 (1941). See also No. 41.

43. Spain and Portugal: Vols. 32 (1909), 35 (1912), 38 (1915–18), 45 (1930). See also No. 41.

44. Switzerland: Vols. 17 (1894), 19 (1896), 21 (1898), 23 (1900), 26 (1903), 29 (1906), 32 (1909), 35 (1912), 38 (1915–18), 43 (1928), 55 (1940).

45. Germany: Vols. 17 (1894), 19 (1896), 21 (1898), 23 (1900), 26 (1903), 29 (1906), 32 (1909), 35 (1912), 41 (1926), 48 (1933), 49 (1934).

46. Denmark: Vols. 17 (1894), 19 (1896), 21 (1898), 23 (1900), 26 (1903), 29 (1906), 32 (1909), 35 (1912), 38 (1915–18), 46 (1931), 56 (1941).
47. Norway and Sweden: Vols. 17 (1894), 19 (1896), 21 (1898), 23 (1900), 26 (1903); including Finland: 44 (1929), 56 (1941).
48. Norway: Vol. 35 (1912). See also No. 47.
49. Sweden: Vols. 32 (1909), 35 (1912), 38 (1915–18). See also No. 47.
50. Baltic States: Vols. 43 (1928), 51 (1936).
51. Poland: Vols. 43 (1928), 51 (1936).
52. Austria-Hungary: Vols. 17 (1894), 19 (1896), 23 (1900), 26 (1903), 29 (1906), 32 (1909), 35 (1912). See also Nos. 53 and 54.
53. Austria: Vol. 45 (1930), 58 (1943). See also No. 52.
54. Hungary: Vol. 43 (1928). See also No. 52.
55. Czechoslovakia: Vol. 43 (1928), 55 (1940).
56. Rumania: Vols. 26 (1903), 29 (1906), 32 (1909), 35 (1912), 43 (1928), 53 (1938, Pt. 2). See also No. 41.
57. Balkan peninsula: Vols. 32 (1909), 35 (1912). See also No. 41.
58. Bulgaria: Vol. 48 (1933). See also Nos. 41 and 57.
59. Yugoslavia: Vol. 44 (1929). See also Nos. 41 and 57.
60. Russia in Europe: Vols. 17 (1894), 29 (1906), 35 (1912), 38 (1915–18), 43 (1928).
61. Africa: Vols. 9 (1882), 10 (1884), 12 (1888), 14 (1890–91), 16 (1893), 18 (1895), 20 (1897), 22 (1899), 25 (1902), 27 (1904), 30 (1907), 32 (1909), 36 (1913), 39 (1919–23), 43 (1928).
62. Asia: Vols. 9 (1882), 10 (1884), 12 (1888), 14 (1890–91), 16 (1893), 18 (1895).
63. Asia, without Russian Asia: Vols. 20 (1897), 22 (1899), 25 (1902), 32 (1909), 37 (1914), 42 (1927), 43 (1928), 47 (1932), 52 (1937).
64. Russian Asia: Vols. 27 (1904), 37 (1914), 41 (1926), 53 (1938, Pt. 2), 54 (1939, Pts. 1, 2).
65. Western Asia: Vol. 47 (1932).
66. Eastern Asia: Vols. 42 (1927), 53 (1938, Pt. 1).
67. India: Vol. 42 (1927).
68. Southeastern Asia and Indonesia: Vol. 42 (1927), 57 (1942).
69. Japan: Vols. 43 (1928), 54 (1939, Pt. 2).
70. Australia and Oceania: Vols. 10 (1884), 14 (1890–91), 16 (1893), 18 (1895), 20 (1897), 22 (1899), 25 (1902), 27 (1904), 30 (1907), 32 (1909), 42 (1927), 53 (1938, Pt. 2).
71. Polar regions: Vols. 9 (1882), 10 (1884), 14 (1890–91), 18 (1895), 21 (1898), 27 (1904), 32 (1909), 36 (1913); north polar regions: 44 (1929), 47 (1932); south polar regions: 48 (1933).

TOPOGRAPHIC MAPS (see also No. 5)
72. General review: Vol. 1 (1866).

73. Topographic maps: Vol. 40 (1924–25); Europe: 1 (1866), 4 (1872), 12 (1888), 14 (1890–91).
74. Key maps of the more important topographic maps of Europe and other regions: Vols. 12 (1888), 14 (1890–91), 17 (1894), 19 (1896), 21 (1898), 23 (1900), 25 (1902), 29 (1906), 32 (1909).

Bibliographie géographique internationale
   Paris, published for Association de Géographes Français by Armand Colin. 1891–. Normally annual. First 2 issues appeared in *Annales de géographie* [694], 1891–1892, 1892–1893; Nos. 3–24, 1893–1914 (entitled successively *Bibliographie de l'année . . . , Bibliographie annuelle,* and *Bibliographie géographique annuelle*), issued as separate numbers of the *Annales de géographie;* Nos. 25–49, 1915–1939, issued under the direction of Elicio Colin; Nos. 25–29, 1915–1931, entitled *Bibliographie géographique;* Nos. 30–49, 1931–1939, with present title. Publication is to be resumed.   [7]
   The most convenient, comprehensive, and in many respects the best of all current geog. bibls. Lists titles of books, period. arts., and maps that have appeared during the year. Divided into 2 parts, general and regional. The general part includes sections on the history of geog., mathematical geog. and cartography, physical geog. (*géographie naturelle*) (with subdivisions dealing with geophysics, geology, geomorphology, meteorology and climatology, hydrography, oceanography, and biogeography), and human geog., including anthropology, political geog., economic geog., and colonization. The material included (in most recent numbers, usually between 2800–2900 items a year) is critically selected and to most of the items brief signed critiques are appended. Auth. index.
   After 1923, when the coöperation of the American Geographical Society was first secured, the list of foreign collaborating institutions grew, so that at the time of suspension of publ. it included the American Geographical Society, Comitato Geografico Nazionale Italiano, Deutscher Geographentag, Royal Geographical Society (London), Société Belge d'Études Géographiques, and Société Royale de Géographie d'Egypte.
   In order to bring together conveniently the material in the *Bibliographie géographique* bearing on related subjs., the American Geographical Society has cut apart and reassembled Nos. 33–48 (1923–1938) and bound them in 21 vols. devoted each to a particular topic or region, making in effect a cl. retrospective bibl. of geog. for the sixteen years covered.

Recent Geographical Literature, Maps, and Photographs Added to the Society's Collections: Supplement to the Geographical Journal.
London, Royal Geog. Soc. 1918–1940. Irregularly (usually 4 numbers a year). Index: Vols. 1–4, 1918–1932, published in 1936.   [8]
Succeeds lists published prior to 1918 in *Geog. Journ.* [625]. By regions, with a genl. section and a section for new maps and new photographs; *ca.* 500 entries per issue. Lack of auth. index is a drawback.

Current Geographical Publications: Additions to the Research Catalogue of the American Geographical Society
New York, Amer. Geog. Soc., 1938–. Monthly (except July and August). Mimeographed.   [9]
Covers books, maps, and period. arts. currently acquired by the American Geographical Society. Each issue, which contains some 600 refs., includes a section on genl. works, cl. by topics and arranged system., followed by a section on regional works cl. and arranged by regions; no anns.; topical index (system.); auth. index. Since October, 1940, this work has been accompanied by a photograph supplement.

Bibliotheca Geographica: Jahresbibliographie der geographischen Literatur
Gesellschaft für Erdkunde zu Berlin. 1895–1917 (the first 18 volumes cover the period from 1891 to 1910; the last (Vol. 19), covering 1911–1912, was published in 1917).   [10]
Superseded annual summaries of geog. literature published in the *Zeitschrift* of the Berlin Geographical Society [716] during the period 1853 to 1890. While of exceptional value because of the number of items listed, no critical notes are included. Primarily by regions with a genl. topical section; auth. index.

Petermanns geographische Mitteilungen   [11]
The sections devoted to geog. literature and especially to maps in *Petermanns Mitteilungen* [717], particularly during the period before the Second World War, were of exceptional value as a bibl. tool. The bibls. are especially strong for current German material.

RETROSPECTIVE BIBLIOGRAPHIES AND PUBLISHED LIBRARY CATALOGUES

Survey of general works of a geographical nature, bibliographies, periodicals, reports of progress in research, encyclopedias and dictionaries, general and regional manuals, works on maps and atlases, etc.: Wagner [60], 1938, pp. 1–27, 61–64.—See also [1156]. [12]

Migliorini, Elio. *Guida bibliografica allo studio della geografia.* Naples, Pironti, 1945. 265 pp. (*Guide bibliografiche Pironti,* No. 1.) [13]
A work somewhat comparable to the present volume, in which relatively more attention is given to the critical discussion of various manuals of genl. and systematic geog. and less to genl. non-geog. aids, geog. bibls. and periodicals. Deals primarily with works in western European languages that have appeared during the past quarter century. Chaps. 1–10: refs. followed by running discussions of geog. bibls., periodicals, manuals, world atlases, world maps, dictionaries, statistical annuals, etc.; Chap. 11: unannotated refs. to regional studies. Does not cover hist. geog. or hist. of geog. and exploration. 938 entries.

Wright, J. K. "Geography and the Study of Foreign Affairs," *Foreign Affairs,* 17 (1938), 153–163. [14]
A short art. with ann. refs. covering outstanding bibl. aids, geog. periods., maps, atlases, gazetteers, and regional geographies.

Hassert, Kurt. *Einführung in die geographische Literatur: Ein Wegweiser für Anfänger.* Dresden, Zahn und Jaensch, 1932. 89 pp. (*Dresdner geographische Studien* [753], No. 3.) [15]
Bibl. aids, geog. periods., gazetteers, lexicons, statistical publs., textbooks, and compendiums are evaluated, but there are no suggestions for regional material or for cartography and atlases. A serviceable introduction to German material within these limits.

Dinse, Paul. *Katalog der Bibliothek der Gesellschaft für Erdkunde zu Berlin.* Berlin, Mittler, 1903. 952 pp. [16]
Covers books, separates, important component parts of larger works, and important period. arts., but not maps. Certain works dealing with more than one subj. are listed in different places. Contains a list of periodicals and an appendix on linguistic works and various nongeog. aids to geog. research. Main catalogue divided in two parts: (1) genl. geog. literature, cl. by topics, incl. bibls., hist. of geog. and discovery, hist. of cartography, methodology, encyclopedic works, series, physical geog., biogeog., anthropogeog., historical geog., geog. names and transcription; (2) literature of particular regions and seas, cl. by regions; auth. index.
The principle of cl. adapted for this great catalogue provided the model for the cl. of the materials in the record section of the *Geog.*

Review [476], in the Research Catalogue of the American Geographical Society [20], and in the present vol.

Mill, H. R. *Catalogue of the Library of the Royal Geographical Society.* London, Murray, 1895. 841 pp. [17]
Somewhat less serviceable than Dinse [16], 1903, because cl. is alph. by auths. and there is no subj. index. Does not cover period. arts. or maps. Three appendixes contain lists of "Collections of Voyages and Travels," of "Government, Anonymous, and Other Miscellaneous Publications" (place cl.), and of periodicals. In 1871 the Society published a *Classified Catalogue of the Library of the Royal Geographical Society* (478 pp.) (alph. by regions). On this and earlier catalogues of the Royal Geographical Society collections, see the preface to the catalogue of 1895. See also [21].

Cardon, F. *Pubblicazioni geografiche stampate in Italia fra il 1800 e il 1890: saggio di catalogo . . . pubblicato in occasione del Primo Congresso Geografico Italiano (Genova, 1892).* (Societá Geografica Italiana.) Rome, 1892. 330 pp. [18]

Koner, Wilhelm. *Repertorium über die vom Jahre 1800 bis zum Jahre 1850 in akademischen Abhandlungen, Gesellschaftsschriften, und wissenschaftlichen Journalen auf dem Gebiete der Geographie, Reisen, Ethnographie und Statistik erschienenen Aufsätze.* Berlin, 1854. 308 pp. (Koner's *Repertorium auf dem Gebiete der Geschichte und ihrer Hülfswissenschaften,* Vol. 2, Pt. 2.) [19]
Comprehensive guide to geog. periods. of the first half of the 19th century.

CARD CATALOGUES OF
LARGE GEOGRAPHICAL LIBRARIES

American Geographical Society Library [20]
The following card catalogues are maintained: (1) a dictionary catalogue covering all books and a selection of outstanding period. arts.; (2) a dictionary catalogue of maps (see [106]); (3) a checklist of periodicals (by countries and cities); (4) "Research Catalogue," covering material received since 1923 (see [9, 16]). The last is system. arranged and divided into two main parts: (a) "general," covering nonregional material, cl. by topics; and (b) "regional," covering regional material, cl. by regions and subcl. by topics within each region. There are also an auth. index providing refs. to the cards in the 2 main parts of the Research Catalogue, a topical index (system.) providing

refs. to the cards in the regional part, and an alph. key to the topical hdgs. The Society's collections of photographs, lantern slides, and newspaper clippings are arranged as in the Research Catalogue. The Society's system of cl. as employed in the Research Catalogue, with modifications and improvements, is explained by S. W. Boggs, "Library Classification and Cataloging of Geographic Material," *Annals Assn. of Amer. Geogs.* [479], 27 (1937), 49–93. See also *Geog. Rev.* [476], 27 (1937), 688–689.

Royal Geographical Society [London] Library [21]
A large card catalogue cl. by subjs. and incl. refs. to papers in serials. See also [17].

BIBLIOGRAPHIES OF WORKS
FOR TEACHERS OF GEOGRAPHY

(For works on geographical education see [187–189].)

Illinois Bulletin of Geography, issued in mimeographed form by the Illinois Chapter of the National Council of Geography Teachers, contains (Vol. 7, No. 4, May, 1945) a brief "Bibliography of High School Geography" (9 pp.).—On certain recent American college textbooks in geography: reviews by E. G. R. Taylor (*Geog. Rev.* [476], 27 (1937), 129–135) and J. O. M. Broek (*ibid.,* 31 (1941), 663–674; in form of a dialogue). [22]

Jensen, J. G., and Marion I. Wright. *Bibliography of the Best References for the Study of Geography.* (Rhode Island College of Education.) Providence, R.I., 1945. 31 mimeo. pp. [23]
Designed especially for "our students, our associates in the public schools, and other adults wishing to develop understanding and knowledge of geography." By topics and regions; ann.

[Clark University: School of Geography.] *Bibliography of Textbooks and Reference Books on Geography.* [Worcester, Mass., 1939?] 6 pts. Unnumbered mimeo. pages. [24]
The 6 parts cover works: (1) for elementary and junior high schools; (2) for senior high schools; (3) for teachers' colleges, liberal arts colleges, universities, and graduate schools; (4) on the teaching of geog.; (5) for use in adult education; and (6) in magazines and periodicals. See also *Supplementary Readings in Geography for Elementary and Secondary Schools,* Clark University, April, 1939. In 2 parts, for Grades 1–2 and 2–3 respectively.

Wright, Grace S. *Government Publications of Use to Teachers of Geography and Science.* Washington, 1938. 16 pp. (U.S. Office of Education, Leaflet No. 31.)   [25]
Alph. by places and topics.

Geographical Books: A Bibliographical List of University and School Text and Reference Books Prepared for the Use of Teachers and Students. Cambridge, England, Heffer, 1934. 86 pp.   [26]
A selection showing "considerable knowledge and judgment" (*Geography* [629], 19 (1934), 79).

Forsaith, D. M. *A Handbook for Geography Teachers.* London, Methuen, 1932. 348 pp. (*Goldsmith's College Handbooks for Teachers*, No. 2.)   [27]
Although "concerned with the teaching of geography to children under fifteen," contains a wealth of refs. of value to mature students. No index. Consists in the main of ann. bibls. Six main parts: (1) syllabuses and schemes of work; (2) indoor geog.—the geog. room (globes, maps, pictures, lantern slides, etc.); (3) outdoor geog.; (4) book lists (by areas and topics, incl. the various branches of geog., map reading and map making, atlases, periodicals, geog. in literature, etc.); (5) the teaching of international geog.; (6) geog. and the League of Nations.

Booth, Mary J. *Material on Geography, Including Commercial Products, Industries, Transportation and Educational Exhibits Which May Be Obtained Free or at Small Cost.* 5th rev. ed. Charleston, Ill., 1931. 108 pp.   [28]
Main list alph. by places and topics.

Mill, H. R. *Guide to Geographical Books and Appliances: The Second Edition of "Hints to Teachers and Students on the Choice of Geographical Books for Reference and Reading."* London, Philip, 1910. 215 pp.   [29]
The object of this work, as stated in the preface of the first edition, 1897, was "to place before teachers and students a selection of the best available books on geography as an educational subject, and on different parts of the world." Cl. by places and topics.

BOOKS OF TRAVEL, FICTION, DISSERTATIONS

On bibliographical sources for geographical fiction: note in *Geog. Rev.* [476], 28 (1938), 499–501   [30]

Holmes, U. T., Jr. *Books of Travel.* Rev. ed. 1931. 54 pp. (*Univ. of North Carolina Extension Bull.,* Vol. 10, No. 10.)   [31]

Brief selective lists of popular books of travel in English with reviews of some of the more outstanding among them. Place cl.

Wharton, Miss D. *Short List of Novels and Literary Works of Geographic Interest.* Prepared by the Leeds Branch of the Geographical Association, with the coöperation of the other Yorkshire branches. Aberystwyth, n.d., [1921?]. 22 pp.   [32]

———

Lists of dissertations in geography accepted by universities in the United States for the degree of Ph.D. will be found in *Annals Assn. Amer. Geogs.* [479], 25 (1935), 211–237, and 36 (1946), 215–247. See also p. 35, n. 34, above.   [33]

GEOGRAPHICAL INSTITUTIONS

GEOGRAPHICAL CONGRESSES

International geographical congresses are listed below. For reports on national geographical congresses or their equivalents held in France, Germany, and Italy see [690], [715], and [840].

International Geographical Congresses   [34]
Each report has been edited in the country in which the congress was held and in the language of that country. Individual monographs in each report are in various languages. The following list shows places and dates of the congresses and in parentheses the dates of publ. of the reports (also place of publ. where not the same as that of a congress). See also above, p. 19, n. 10.

First, Antwerp, 1871 (1872); Second, Paris, 1875 (1878); Third, Venice, 1881 (Rome, 1882); Fourth, Paris, 1889 (1890); Fifth, Berne, 1891 (1892); Sixth, London, 1895 (1896); Seventh, Berlin, 1899 (1901); Eighth, in the United States, 1904 (Washington, 1905); Ninth, Geneva, 1908 (1909); Tenth, Rome, 1913 (1915); Eleventh, Cairo, 1925; Twelfth, Cambridge, 1928; Thirteenth, Paris, 1931; Fourteenth, Warsaw, 1934; Fifteenth, Amsterdam, 1938.

GEOGRAPHICAL SOCIETIES, UNIVERSITY DEPARTMENTS, INSTITUTES, AND ASSOCIATIONS

Many of the geographical societies of the world are semiscientific, semisocial organizations whose membership consists primarily of persons interested in travel and geography—merchants, professional men, teachers, government officials, army and navy officers, etc., living in or near a particular city. These societies may be called centralized. Lectures are held, rooms and sometimes a building containing a library and map collection are often maintained, and a periodical is usually published. A few of the larger societies, however, originated as, or have developed into, national institutions, possessing large libraries and employing professional staffs engaged in geographical work—notably the Royal Geographical Society (London), the American Geographical Society (New York), the National Geographic Society (Washington).

Another type of centralized geographical organization is the university or college department of geography. As a very rough indication of the stronger departments, the following list shows the colleges and universities in the United States having two or more members of the Association of American Geographers on their faculties (June, 1945) (see also below, p. 277): Carleton Coll. (2); Clark Univ. (3); Columbia Univ. (5); Harvard Univ. (2); Indiana Univ. (3); Northwestern Univ. (3); Ohio State Univ. (8); Stanford Univ. (2); Syracuse Univ. (3); Univ. of California (8); Univ. of Chicago (7); Univ. of Cincinnati (3); Univ. of Maryland (2); Univ. of Michigan (6); Univ. of Minnesota (3); Univ. of Missouri (2); Univ. of Pennsylvania (3); Univ. of Washington (Seattle) (2); Univ. of Wisconsin (5).

Associated with certain university departments, or sometimes more or less independently established by an individual professor, are geographical institutes. These are more common in continental Europe than in the United States. Many of them issue series or serial publications.

As distinguished from these centralized organizations there are certain decentralized associations of professional geographers and teachers of geography, such as the Association of American Geographers, Association de Géographes Français, Geographical Association (Great Britain), National Council of Geography Teachers (U.S.), and the newly organized American Society for Professional Geographers. Most of these are national in scope; they hold meetings in different places and publish periodicals of a more specialized type than those of the majority of the centralized geographical societies.

Short reports on the annual meetings of the Association of American Geographers, often with brief summaries of papers read, have been published in the April number of the *Geographical Review* [476] each year. Abstracts of the papers are published in the *Annals of the Association of American Geographers* [479].

Yet another type of geographical organization is the section or division devoted to geography of a large scientific body, such as the Section E, Geology and Geography, of the American Association for the Advancement of Science, the Division of Geology and Geography of the National Research Council (U.S.), and Section E, Geography, of the British Association for the Advancement of Science.

For data on these various organizations see *Minerva* and *Index Generalis* (notes 7, 8, on pp. 18, 19 above), *Geographen Kalender* [725]; also [6, Nos. 29, 30], [36, 106] and references to the serial publications of geographical societies in Part III below.   [35]

Sparn, Enrique. "Cronologia, diferenciación, numero de socios y distribución de las sociedades de geografía," *Boletín de la Academia Nacional de Ciencias,* Córdoba, Argentina, 32 (1935), 323–336. [36]
Statistical information largely from *Minerva,* Vol. 30, 1930; Vol. 31, 1933, and *Index Generalis,* 1933 (see above p. 18). Accompanied by 3 sketch maps showing distribution of geog. societies.

GEOGRAPHICAL LIBRARIES AND MUSEUMS

(See also above, pp. 22–26.)

According to the survey recently made by the American Library Association (see above, p. 23, note 20) the following American libraries contain important collections of geographical publications (see also below, p. 276):

University libraries: California (Latin America, Pacific countries); Chicago (United States, Europe, Latin America, East Asia, India); Clark; Columbia; Harvard; Louisiana; Michigan (United States, Latin America, East Asia); Texas (Latin America); Washington at Seattle (Northwest); Yale. *Public library:* New York (historical, relating to America). *Other libraries:* American Geographical Society; Explorers' Club; Library of Congress; National Geographic Society; Pan American Union (Latin America); United States Department of Commerce. See also E. T. Platt, "Books and

National Defense: A Brief Survey of Some Library Resources of Geographical Pertinence," *Geog. Rev.* [476] 31 (1941), 264–271. [37]

According to Sparn [36], 1935, p. 335, in the early 1930's there were 21 geographical societies in the world with libraries containing more than 20,000 volumes each, as follows: Paris (300,000 vols.); Russian State (formerly Imperial) (180,000); American (N.Y.) (97,500); Royal (London) (80,000); Berlin (60,000); Italian (Rome) (60,000); Vienna (50,000); Far East (Khabarovsk) (50,000); Lisbon (40,000); Madrid (35,000); Brisbane (30,000); Vladivostok (26,500); Marseille (25,000); Nancy (25,000); Tashkent (23,000); Edinburgh (22,000); Dresden, Leipzig, Paris (Géogr. Commerciale), 20,000 each. The geographical societies having the largest map collections were London (170,000 maps); American (N.Y.) (85,700); Paris (60,000); Berlin (20,000). These figures, derived from reports made by the societies themselves to *Minerva* and *Index Generalis,* are unreliable and may not be comparable but are the only data available.

On Jan. 1, 1946, the Library of the American Geographical Society contained approximately 116,600 volumes of books and periodicals; 23,000 pamphlets; 133,400 maps; 2,300 atlases (*Geog. Rev.* [476], 36 (1946), 317). [38]

———

Reinhard, Rudolf. "The Museum of Regional Geography in Leipzig," *Geog. Rev.* [476], 24 (1934), 219–231. [39]

Semenov-Tian-Shansky, Benjamin. "The Geographical Museum," *Geog. Rev.* [476], 19 (1929), 642–648. [40]
Deals with a museum in Leningrad.

GEOGRAPHICAL PERIODICALS AND SERIES

GEOGRAPHICAL PERIODICALS

The scientific and scholarly quality of geographical periodicals (the term "periodical" should be understood as embracing also society transactions, etc.) varies widely. Some are critical instruments of research, including articles, maps, bibliographies, and book reviews of merit and dealing with subjects of world-wide scope. Others of equally high quality specialize in a particular region or group of topics. Still others, though their nominal aim is to treat the geography of the world as a whole, are of little significance except for the special data they may furnish regarding their respective regions (see p. 40). Finally there are not a few so-called geographical periodicals so given over to antiquarian and historical studies or to the "travelogue" type of article as to be of little or no use to the professional geographer.

The great majority of geographical periodicals are issued by geographical institutions of the types indicated in [35]. A few, including some of the more scholarly as well as the more popular, are or have been published by commercial houses. Many of the older geographical-society periodicals present little more than the transactions of meetings, texts of the lectures, and occasional independent papers. During the nineteenth century their pages were often filled largely with data on current explorations. With the growth of geography as a profession a new type of periodical made its appearance, in which attention was focused on the scientific, academic, or educational phases of the subject. Most of the serial publications of geographical institutes connected with universities are of this type, and, in addition, in nearly every large country one or more strictly professional periodicals, national or even international in the scope of their contributors, made their appearance between 1890 and 1920, paralleling the periodicals of the nation-wide geographical societies.

Thus we have in France the newer *Annales de géographie* [694] established in 1891 and the older *Bulletin* of the Paris Geographical Society [691] dating from 1822; in Italy the *Rivista geografica italiana* [843], 1894–, and the *Bolletino* of the Italian Geographical Society [841], 1868–; in Russia *Zemlevedenie* [897], 1894–, and the journals of the Russian Geographical Society [895], 1865–; in Germany *Geographische Zeitschrift* [719], 1895–, and the *Zeitschrift* of the Berlin Geographical Society [716], 1839–; in Great Britain *Geography* (formerly the *Geographical Teacher*) [629], 1901–, and the *Journal of the Royal Geographical Society* [625], 1830–; and in the United States the *Annals of the Association of American Geographers* [479], 1911–, and the *Bulletin of the American Geographical Society* [476], 1852–. In all of these instances the new professional periodical was issued apart from the national geographical societies, either as a more or less independent venture (e.g., *Geographische Zeitschrift* and *Zemlevedenīe*) or by a nation-wide association of professional geographers. In Sweden and the United States, however, geographical societies have taken the initiative in producing professional journals. *Geografiska Annaler* [648] was established in 1919 by the Swedish Society for Anthropology and Geography as an international organ of scientific geography along with the same society's older journal [647], 1881–, and the *Geographical Review* [476] superseded the *Bulletin* as the organ of the American Geographical Society in 1916. It should be noted that many of the journals of nation-wide geographical societies, though less strictly professional than the newer periodicals, nevertheless contain material of the highest scientific and scholarly quality. This is notably true of the *Geographical Journal,* the *Zeitschrift* of the Berlin, the *Izvestiia* of the Russian, and the *Bolletino* of the Italian geographical societies. A few geographical societies issue illustrated popular magazines (one of which, the *National Geographic Magazine* [477], has an extremely wide circulation), though other magazines of this type (e.g., [630], [651], [682]) are due to the enterprise of commercial publishers. On the other hand one commercially published geographical periodical, *Petermanns Mitteilungen* [717], long stood in the first rank among the scholarly instruments of geographical research.

It should be noted that the sixteen titles given in the preceding paragraph are those of the world's outstanding geographical periodicals of broad scope.

Most of the periodicals listed in this volume are geographical either in the strict sense of the term (see p. 30) or at least are nominally so. Probably the majority, though by no means all, of the periodicals of geographical societies and geographical institutes are included as well as many others which contain the word "geographical" in their titles. A few nonregional periodicals devoted to special branches of geography are listed in Part II; general and regional geographical periodicals and a few other regional periodicals that deal with large regions as a whole are listed in Part III. Only such nongeographical periodicals are listed (e.g., historical, ethnographic) as contain notable material of geographical value.

In order to find geographical literature in the periodicals of the first half of the 19th century the reader should make use of Koner [19], 1854. During the period since 1850 the number of periodicals has grown so great that the compilation of another work like that of Koner has never been attempted. To find geographical material in the serials of the last ninety-five years we must look to the retrospective and current bibliographies that contain references to periodical articles (see above, pp. 51–60).

Among American libraries especially strong in their collections of geographical periodicals, that of the American Geographical Society probably ranks first.

In *Geographen-Kalcnder* [725], 1914, pp. 353–414, may be found an alphabetical list of periodicals dealing with geography and related sciences. Here and on pp. 69–350 under the name of the publishing institution, data are given regarding editors, frequency of publication, and date when publication was begun. A list, arranged regionally, of specifically geographical periodicals is included in *Geographisches Jahrbuch* [6], 1909, pp. 419–429. See also Wagner [60], 1938, pp. 6–8. [41]

On indexes to *Geographical Journal* [625], *Petermanns Mitteilungen* [717], *Scottish Geographical Magazine* [626], *Geographical Review* [476], *National Geographic Magazine* [477]: note in *Geog. Rev.* [476], 27 (1937), 336–337. [42]

GEOGRAPHICAL SERIES

The following references cover a selection only of the more important general geographical series. Others are listed in Part III under the countries in which they are published.

SERIES PUBLISHED BY GEOGRAPHICAL SOCIETIES AND OTHER SCIENTIFIC INSTITUTIONS

Royal Geographical Society, London. (a) *Supplementary Papers.* 1886–1893 4 vols. (b) *Technical Series.* 1920–[1933]. Nos. 1–[5]. [43]
(a) consists of papers "too long or too technical for inclusion in the monthly publication" of the Society. (b) deals primarily with surveying and map making. The Society has also published several bibls., maps, monographs, and educational works that do not form parts of specific series. See H. R. Mill, *The Record of the Royal Geographical Society, 1830–1930,* London, Royal Geog. Soc., 1930.

University of California Publications in Geography. 9 vols. 1913–
[1944]. [44]
Each vol. includes monographs numbered serially.

American Geographical Society, New York. (a) *Special Publications.*
1915–[1945]. Nos. 1–[29]. (b) *Research Series.* 1922–[1947].
Nos. 1–[22]. [45]
These 2 series are essentially similar in nature, (a) being issued in a
larger format than (b) to permit the adequate reproduction of maps
and photographs. Both series comprise for the most part monographs
presenting the results of original research. The numbers are bound
volumes except for 2 slip cases each containing pamphlets and maps
in (a) and 5 pamphlets in (b).
The Society also publishes other series of more limited scope. See
American Geographical Society: Books and Maps [1946], a leaflet in
which the publs. of the Society are listed.

National Geographic Society, Washington. *Technical Papers,* con-
stituting 4 series: *Katmai Ser.,* 2 vols., 1923–1924; *Pueblo Bonito
Ser.,* No. 1, 1935; *Stratosphere Ser.,* Nos. 1, 2, 1935–1936; *Solar
Eclipse Ser.,* Nos. 1, 2, 1939–1942. [46]

SERIES PUBLISHED BY COMMERCIAL PUBLISHERS

British

(a) *The Regions of the World,* ed. by H. J. Mackinder. Clarendon
Press, Oxford Univ. Press (published in United States as *Appleton's
World Series*). 6 vols., including such classics in regional geography
as Mackinder's *Britain and the British Seas,* 1902, and D. G. Ho-
garth's *The Nearer East,* 1902. (b) *Cambridge Geographical Series.*
London. 6 vols. (c) *Methuen's Advanced Geographies,* London
(published in the United States by E. P. Dutton without series
title). Some 10 recent works on regional geography. (d) *Harrap's
New Geographical Series,* London. Several advanced textbooks and
manuals of systematic and regional geography. [47]

French

(a) *Collection Armand Colin: Section de géographie.* Paris. (b)
Géographie pour tous. Paris, A. Fayard & Cie. (c) *Géographie
humaine: Collection dirigée par Pierre Deffontaines.* Paris, Galli-
mard. [48]

German

The following series of small volumes contain works by leading authorities on divers aspects of geography and nearly all parts of the world: *Aus Natur und Geisteswelt,* Berlin, Teubner; *Sammlung Göschen,* Berlin and Leipzig, Walter de Gruyter; *Jedermanns Bücherei: Abteilung Erdkunde,* Breslau, Ferdinand Hirt; *Perthes Kleine Völker- und Länderkunde zum Gebrauch im praktischen Leben,* Gotha and Stuttgart, F. A. Perthes. [49]

GENERAL GEOGRAPHICAL MANUALS

In this section references are given to a selection of works which the geographical student may profitably consult in order to obtain a broad initial orientation before undertaking more detailed investigation of almost any geographical topic or region. For additional titles see Wagner [60], 1938, pp. 10–14, and Migliorini [13], 1945, pp. 58–73.

For guidance to sources of statistical information, see the references on p. 38, note 45, above. Especially well conceived and arranged are the geographical statistical tables in the *Encyclopædia Britannica World Atlas* [116], "Geographical Summaries."

During both the First and Second World Wars comprehensive regional studies were made of large parts of the world for intelligence purposes by various branches of the governments and armed forces of the belligerent nations. Geographers collaborated in this work and the various volumes that resulted constitute an important source of geographical information. After the First World War the Geographical Section of the Naval Intelligence Division of the British Admiralty and the Historical Section of the British Foreign Office brought out a series of *Manuals* and *Handbooks,* respectively, on the geography and political and economic conditions of most parts of the world. A comparable series of *Notices* was issued by the French Service Géographique de l'Armée. During the Second World War the Military Government Division of the United States War Department prepared a series of *Civil Affairs Handbooks* in about 180 parts. These have been published as *Army Service Forces Manuals* 350–370 and have been distributed to certain American libraries. They "cover the Axis countries and those occupied by them, presenting a wide range of data about the economic, cultural, and political organization of the various countries. The sections and atlases for the cultural institutions should be noted with particular interest." (*Library Journ.*, 71 (1946), 481). Among similar series that have not as yet been "declassified" but which may ultimately be made available to the public are: (1) the *Joint Army Navy Intelligence Studies* (JANIS) dealing primarily with the Far East; (2) the British *Inter-Service Intelligence Studies* (ISIS) prepared at Oxford, England, and covering more particularly Europe and Africa; and (3) the *Handbooks* prepared by the British Division of Naval Intelligence. [50]

On a selection of German geographical works of comprehensive scope: note in *Geog. Rev.* [476], 24 (1934), 674–676. [51]

COLLABORATIVE WORKS

Géographie universelle. Publiée sous la direction de P. Vidal de la Blache et L. Gallois. 15 "tomes" (some in more than one vol.). Paris, Armand Colin, 1927–[1942]. [52]
Several leading French geographers have coöperated in the production of this superb set of regional geogs., notable for the skilful selection and arrangement of the materials, the clarity and distinction of the style, and the fine quality of the maps and photographs. While comprehensive and based on careful scholarship, the vols. are intended for the general reader as well as the specialist. Bibl. refs. are given. All parts of the world have been covered to date.

Klute, Fritz, ed. *Handbuch der geographischen Wissenschaft.* Potsdam, Athenaion, 1930–[1940]. 13 vols. [53]
2 vols. on genl. and 11 on regional geog., by 60-odd contributors. The aim is to render geog. interesting to the general reader, and hence the work is lavishly illustrated with maps, photographs, and colored views (see *Geog. Rev.* [476], 24 (1934), 159–161). Bibl. refs.

Meinardus, Wilhelm, ed. *Allgemeine Länderkunde der Erdteile. . . .* Hannover, Hahn, 1928–[1935]. [54]
Planned as a continuation of Wagner's *Lehrbuch* [59], Vol. 2, to consist of 7 vols. on the genl. geog. of the continents and to reflect Wagner's scholarly spirit. Primarily for the geographer rather than for the general reader. Extremely full bibl. refs. Vols. on North America (1928) and on Australia and Oceania (1931) had appeared before 1939.

Sievers, Wilhelm, and others, eds. *Allgemeine Länderkunde.* Leipzig and Vienna, Bibliographisches Institut, 1891–1933. [55]
The latest editions of the vols. in this series of regional geogs. are: Emil Deckert and F. Machatschek: *Nordamerika,* 1924; Otto Maull: *Deutschland,* 1933; Ludwig Mecking: *Die Polarländer,* 1925; Alfred Philippson: *Europa ausser Deutschland,* 1933; *idem* and Ludwig Neumann: *Europa,* 1906; Wilhelm Sievers: *Asien,* 1904; *idem: Süd- und Mittelamerika,* 1914; *idem* and Fritz Jaeger: *Afrika,* 1928; *idem* and Walter Geisler: *Australien und Ozeanien,* 1930. Each vol. contains bibl. data.

Stanford's Compendium of Geography and Travel. New issue. London, Stanford, 1894–1925. 12 vols. [56]
2 vols. to each continent, of which the most recent are: Europe, 1 (1924), 2 (1925); Africa, 2d ed., 1 (1907), 2 (1904); Asia, 1, 2 (1896); Australasia, 1 (2d ed., 1907), 2 (1894); Central and South America, 2d ed., 1 (1909), 2 (1911); North America, 1 (2d ed., 1915), 2 (1898). Comprehensive if now somewhat old-fashioned series of regional geogs.

Mill, H. R., ed. *The International Geography,* by Seventy Authors. New York, Appleton, 1909. 1,108 pp. (1st ed., London, Newnes, 1898.) [57]
Short monographs on the various parts of the earth, each by an authority; lists of refs. at the end of each contribution.

MANUALS BY INDIVIDUAL AUTHORS

James, P. E. *An Outline of Geography.* Including a Study Guide by C. F. Kohn. Boston, Ginn and Co., 1943. 568 pp. [58]
This distinctive modern textbook contains a short bibl. (pp. 419–426) of refs. "selected with a view not only to covering, as far as possible, the various parts of the world but also to illustrating the various methods of regional description and interpretation." The list is notable for its refs. to significant recent period. arts.

Wagner, Hermann. *Lehrbuch der Geographie.* Vol. 1. *Allgemeine Erdkunde.* 11th ed., Part 1 (see [60]). 10th ed., Part 2. *Physikalische Geographie.* Hannover, Hahn, 1922. Pp. 257–660. Part 3. *Biologische Geographie; Anthropogeographie.* 1923. Pp. 661–1100. Vol. 2. *Länderkunde von Europa,* by H. Wagner and Max Friederichsen. 6th ed. Part 1. *Allgemeine Länderkunde von Europa.* 1915. 184 pp. (See also [54].) [59]
Professor Wagner was unrivaled in his profound and critical knowledge

of geog. literature and methods. This knowledge is reflected in the great
wealth of refs. provided in the *Lehrbuch*. Each of the main subdivisions
of the work is introduced by a bibl. note, and numerous works dealing
with specific points of detail are cited in footnotes. Succinct critical
comments on many of the publs. cited.

Wagner, Hermann. *Allgemeine Erdkunde.* 11th ed., ed. by Wilhelm
Meinardus. Part 1, *Mathematische Geographie.* Hannover, Hahn,
1938. 394 pp.   [60]
The preceding editions of this work formed part of Wagner's *Lehrbuch*
[59], but the title "Lehrbuch" has been dropped in this edition. The
first 64 pages of this vol. consist of a genl. introduction to the *Allgemeine
Erdkunde,* including an important bibl. guide to the whole field of
geog.

Newbigin, Marion I. *A New Regional Geography of the World.* London, Christopher; New York, Harcourt Brace, 1929. 451 pp.   [61]

Unstead, J. F., and E. G. R. Taylor. *General and Regional Geography
for Students.* 10th ed. London, Philip, 1927. 529 pp.   [62]

Lautensach, Hermann. (a) *Allgemeine Geographie zur Einführung
in die Länderkunde: Ein Handbuch zum Stieler.* Gotha, Justus
Perthes, 1926. 445 pp. (b) *Länderkunde: Ein Handbuch zum
Stieler.* 1926. 843 pp.   [63]
Text designed to accompany Stieler's *Atlas* [123]. Each vol. contains
a bibl., 407 entries in (a), 755 in (b).

Reclus, Elisée. *Nouvelle géographie universelle.* Paris, Hachette,
1878–1894. 19 vols. (English translation ed. by E. G. Ravenstein and
A. H. Keane, New York, Appleton, 1882–1895.)   [64]
Probably the most ambitious attempt ever made by a single writer to
encompass the entire subject of geog. Today largely superseded by
more critical works.

GAZETTEERS AND RELATED WORKS

GENERAL GAZETTEERS AND GLOSSARIES

In this section we deal with gazetteers and related reference works
in which the material is arranged in alphabetical order.
General encyclopedias present a wealth of geographical information.

Since, however, these are adequately discussed in Mudge-Winchell [1], a 39–50, b 13–14, c 15–16, d 13–15, no comment seems needed other than to call attention to the *Enciclopedia italiana di scienze, lettere, ed arti* (1929–1934, 24 vols.) and the *Enciclopedia universal illustrada Europeo-Americana* (Barcelona, 1905–1933, 80 vols. in 81), both of which are especially noteworthy for their geographical articles and maps.

Data on dictionaries of languages will be found in Mudge-Winchell [1], a 51–85, b 15–18, c 17–21, d 15–20.

Gazetteers and related reference works provide information in alphabetical order concerning: (1) specific places, regions, and geographical features; (2) geographical names as such; (3) the meaning of generic terms for geographical features; (4) the meaning of other technical terms used in geographical studies.

Works which provide information regarding (1) *specific places*, etc., are usually called *gazetteers* or *geographical dictionaries*. Normally a gazetteer provides brief, concise accounts of the places, etc., listed in it (see Mudge-Winchell [1], a 322–326), and the gazetteers referred to below [66–73] are of this type. The term "gazetteer," however, is also applied to indexes to atlases or maps, in which no information is furnished other than names and their locations. Military needs have recently stimulated the preparation and publication of "gazetteers" of this kind by such agencies as the Hydrographic Office, the United States Board on Geographical Names, and the Army Map Service in this country, and the Geographical Section of the General Staff (G.S.G.S.) and the Permanent Committee on Geographical Names in Great Britain.

Works dealing with (2), *geographical names as such,* are of two main types: (a) those in which emphasis is laid on the historical or linguistic aspects of such names (for these see Mudge-Winchell [1], a 327–332, b 46–47, c 66, d 60–61; Wagner [60], 1938, pp. 23–24, and below [74]), and (b) those designed primarily as aids to consistency in the spelling of geographical names (see [76–85]).

Generic terms applied to geographical features (3) form integral parts of innumerable place names in all languages (e.g. *Mount* Washington, Iwo *Jima, Wady* Halfa). Hence their meaning must be understood if maps and other geographical works are to be properly prepared and interpreted. Many of these terms have equivalents in nearly all languages, since they apply to features of the landscape found throughout the world (e.g. *hill, valley*); other terms apply to features found only in special environments of limited extent (e.g. *erg, salar*) and consequently are restricted to the languages and dialects spoken where these features occur. Some of these terms, moreover, are of purely local currency among peasants, fishermen, etc., and never find their way into the language dictionaries. A pioneer glossary of such terms is that of Knox [75], 1904. Olson and Whitmarsh [96], 1944, contains brief glossaries for many languages. Recently the Army Map Service has been issuing a comprehensive series of glossaries providing English translations of geographical terms in a large number of languages. While some of these glossaries have been compiled by the Army Map Service itself, the majority are reproduced from pamphlets issued by the G.S.G.S. and the Permanent Committee on Geographical Names for British Official Use (see [78, 79]).

From a good many different languages, geographers, geologists, and other scientists have borrowed as technical terms a large number of generic terms for geographical features (e.g. *esker, cirque, bolson*). They have also coined new terms, or applied new meanings to existing words, in order to designate geographical features for which terms do not exist in popular speech (e.g. *peneplane, monadnock*). Furthermore, the literature of geography and its related fields abounds in (4) *technical or semitechnical terms* that do not apply to geographical features (e.g. *latitude, phytogeography, sequent occupance*). Since many of these terms are not included in the ordinary dictionaries, or, where so included, are not always defined with the precise shades of meaning with which geographers use them, there is a genuine need for an English dictionary of scientific geographical terms, and also for bilingual or multilingual dictionaries or vocabularies giving the precise equivalents of such terms in different languages.

For references to a selection of technical dictionaries and glossaries of geographical value see *Geog. Rev.* [476], 30 (1940), 331–332. See also [6, No. 31].   [65]

Cassell's World Pictorial Gazetteer. Ed. by Sir J. A. Hammerton. London, Cassell [1940]. 1,024 pp.   [66]
Medium-sized, inexpensive gazetteer, illustrated with line cuts in text.

Garollo, Gottardo. *Dizionario geografico universale.* 5th ed. rev. by

Arrigo Lorenzi. Milan, Ulrico Hoepli, 1929–1932. 2 vols. 2,191 pp.
(*Manuali Hoepli.*) [67]
Deals with geog. names and terms.

Nelson's World Gazetteer and Geographical Dictionary. With Map
Supplement by Bartholomew. General editor, J. Gunn. London,
T. Nelson [1931]. 575 pp. [68]
Handy small gazetteer.

Kende, Oskar. *Geographisches Wörterbuch: Allgemeine Erdkunde.*
2d ed. Leipzig and Berlin, B. G. Teubner, 1928. 238 pp. (*Teubners
kleine Fachwörterbücher,* Vol. 8.) [69]
Arts. on places and regions are not included. On the other hand there
are entries on peoples and races and short biographies of eminent geographers and explorers. The most useful features for the English-speaking student are the brief explanations of German technical terms, particularly in the fields of physical and mathematical geog. Reviewed in
Geog. Rev. [476], 20 (1930), 175–176.

Banse, Ewald. *Lexikon der Geographie.* 2d ed. Brunswick and Hamburg, Westermann, 1923. 2 vols. Vol. 1, 786 pp.; Vol. 2, 785 pp. [70]
Deals not only with places and regions, as in a gazetteer, but also with
geog. terms, with geog. aspects of such broad topics as agriculture,
ethnology, commerce, etc., and with the lives of geographers and explorers.

Lippincott's New Gazetteer. Philadelphia, Lippincott, 1906 (reprinted
in 1931). 2,053 pp. [71]
Lays special emphasis on the United States. While the picture of the
world is that of 1905, the 1931 reprint includes a conspectus of the
United States Census of 1930.

Longman's Gazetteer of the World. G. G. Chisholm, ed. London,
Longmans, 1902. 1,788 pp. [72]
Especial emphasis on the English-speaking world.

Vivien de St. Martin, Louis. *Nouveau Dictionnaire de géographie
universelle.* Paris, Hachette, 1879–1895. 7 vols. *Supplément,* 1900.
2 vols. [73]
The most elaborate gazetteer in existence—a truly monumental work
containing more than ten times as many words as Banse's *Lexicon* [70]
and nearly three-fifths as many words as does the 14th ed. of the
Encyclopædia Britannica. The articles on the larger regions run to con-

siderable length; those on the United States alone would fill nearly 200 pages if printed as is the present volume. Physical, political, hist., and economic geog., and ethnology are all covered. Many of the articles incorporate tables of statistics—now seldom of value except from the historical point of view. To most of the longer articles bibl. refs. are appended, but these are also out of date. The *Dictionnaire,* however, is still useful for those who wish to obtain facts about small and obscure places—particularly in Europe.

Wright, J. K. "The Study of Place Names: Recent Work and Some Possibilities," *Geog. Rev.* [476], 19 (1929), 140–144. [74]

Knox, Alexander. *Glossary of Geographical and Topographical Terms and of Words of Frequent Occurrence in the Composition of Such Terms and of Place-Names.* London, Edward Stanford, 1904. 472 pp. (*Stanford's Compendium of Geography and Travel,* Supplementary Volume.) [75]
English definitions of terms that occur in a good many different languages. For more recent data of the same type see Olson and Whitmarsh [96], 1944.

SPELLING OF GEOGRAPHICAL NAMES

How to spell geographical names is a perplexing problem that nearly every geographer has to face when preparing manuscripts for publication. Gazetteers and indexes to atlases are not always reliable or consistent. For the benefit of writers in the English language certain approved lists have been prepared by official agencies of the American and British governments [77–81], and for names not included in the lists rules for ascertaining suitable spellings and for transliterating from foreign alphabets have been drawn up [82, 85]. Complete uniformity of practice is not to be expected, and some scholars may find good reasons for not always following the official recommendations. A uniform system, however, should always be followed throughout any one publication. [76]

The United States Board on Geographical Names (formerly United States Geographic Board) is an agency of the United States Government (now in the Department of the Interior) established for the purpose of standardizing the spelling of geographical names in government documents. Two older publications of the Board are listed below [80, 81]. Since Pearl Harbor the activities of the Board have been greatly enlarged and it has issued a large number of pamphlets giving approved spellings and directions for the treatment of names in many different parts of the world. An unpublished index map has been prepared for consultation at the American Geographical Society showing for each country the various publications of the Board in which may be found decisions concerning names in that country. See M. F. Burrill, "Reorganization of the United States Board on Geographical Names," *Geog. Rev.* [476], 35 (1945), 647–652.   [77]

Permanent Committee on Geographical Names for British Official Use. [Lists of names.] 1921–.   [78]
For this committee the Royal Geographical Society has published in a series of pamphlets (listed in Mudge-Winchell [1], a 327, b 46, c 66) genl. lists of approved European, Asiatic, African and Oceanic (Pacific) names, and names for various countries and regions in the Eastern Hemisphere.

Permanent Committee on Geographical Names. *Rules for the Spelling of Geographical Names (Termed the R.G.S. II System)*. London, 1932. 4 pp. (See also [85].)   [79]

Sixth Report of the United States Geographic Board, 1890 to 1932. Washington, 1933. 834 pp.   [80]
"In effect a handbook of geographic names. . . . The outstanding characteristic . . . is its scholarly quality. . . . Of the 25,000 names that the list itself contains, some 16,000 relate to the continental United States, 2,000 to Alaska, 3,000 to the Philippines, 900 to Hawaii, 300 to Puerto Rico and the Virgin Islands of the United States, . . . and 2,500 to foreign countries" (*Geog. Rev.* [476], 25 (1935), 150–151). Introductory chapter on the characteristics of geog. names, problems involved in naming geog. features, names in the United States and its possessions, foreign geog. names; and on the United States Geographic Board. The data on foreign names in [81] are incorporated in this report. Supplementary pamphlets entitled *Decisions of the United States Board on Geographical Names* (which superseded the Geographic Board in 1934) have been issued. It is understood that another comprehensive volume of decisions rendered since 1932 will be issued. See [77].

United States Geographic Board. *First Report on Foreign Geographic Names*. Washington, 1932. 113 pp. (See also [80].)   [81]

TRANSLITERATION

For data on systems of transliteration covering certain other languages besides the ones listed under [85], see Sarton's *History of Science*

[147], Vol. 1, pp. 46–51. The Library of Congress has published in convenient form on catalogue cards rules for transliterating certain alphabets. [82]

"The Names of Countries Used by Other Countries," *Geog. Journal* [625], 85 (1935), 458–462. [83]

"The Spelling of Foreign Geographical Names," *Geog. Rev.* [476], 23 (1933), 147–149. [84]

Gleichen, Edward, and J. H. Reynolds. *Alphabets of Foreign Languages Transcribed into English According to the R.G.S. II System.* 2d ed. London, Royal Geog. Soc., 1933. 76 pp. (*R.G.S. Technical Series,* No. 2.) [85]

Rules for transliteration from some 56 languages or alphabets classed under the following headings: Teutonic, Keltic, Romance, Slavonic, Baltic, Other Indo-European, Caucasian, Finno-Ugrian, Arabic character (includes Arabic, Turki, Persian and Pashtu, Malay, Turkish), Semitic, and Miscellaneous (Basque, Cape Dutch, Maltese). Does not cover Chinese, Japanese, Hindustani, Javanese, etc. (see [82]).

TRAVELERS' MANUALS AND GUIDEBOOKS

TRAVELERS' MANUALS

(See also above, pp. 19, 20.)

Hints to Travellers. 11th ed. Vol. 1, *Survey and Field Astronomy,* by E. A. Reeves. Royal Geographical Society, London, 1935. 448 pp. Vol. 2, *Organization and Equipment, Scientific Observation, Health, Sickness and Injury.* Ed. by the Secretary with the help of many travellers. 1938. 487 pp. [86]

Vol. 2 is intended "principally for the use of travellers with their own organization of transport and supply, camp equipment, and followers." The several chapters deal with such practical problems as health, injury, etc., preparation at home, food and drink, clothing, transport, photography, and with scientific observations in the fields of meteorology, vegetation, geology, natural history, anthropology, and "antiquities" (not geog. as such). The subj. matter is arranged by topics and there are almost no regional entries in the index. Reviewed in *Geog. Rev.* [476], 29 (1939), 348–349.

Brouwer, H. S. ed. *Practical Hints to Scientific Travelers.* The Hague, Martinus Nijhoff, 1925–1929. 6 vols. Distributed in the United

States by the American Geographical Society, New York. [87]
"The principal object of this book is to compile the experiences of
scientific explorers in different countries so that these experiences are
easily accessible. . . . Many travellers, although well equipped with
regard to scientific work, are only poorly informed as to the things of
everyday life as well as to the customs and manners which prevail
among white and colored peoples in distant regions. Yet this knowledge
is hardly less necessary for the success of a journey." Arts. by different
authorities on travel in the following regions: Vol. 1, Netherlands East
Indies, South and East Africa, Philippines; Vol. 2, Polar regions, Spits-
bergen, Novaya Zemlya, Greenland, Turkestan; Vol. 3, Mexico, Indo-
China, India and Burma, New Zealand, New Guinea, Morocco; Vol. 4,
Egypt, Angola, Australia, Antarctica, Venezuela, Haiti; Vol. 5, Ecuador,
Eastern Congo, Northern Manchuria, Amur coastal region of far-
eastern Asia, and Malay Peninsula; Vol. 6, Canada, Argentine Republic,
Oceania, Madagascar, Tropical West Africa.

GUIDEBOOKS

The majority of guidebooks are prepared for tourists and consequently present information concerning matters in which tourists
are interested: "the art museums, collections, etc., of any given place,
its architectural and historical monuments, scenic features, railroad
and other communications, hotels, literary and historical associations, etc." (Mudge-Winchell [1], a 339–340, b 48, d 62). The
better guidebooks also may include general information on physical
and human geography, together with maps and town plans. Such
guidebooks fall into two main categories: (1) those issued as parts
of comprehensive series according to generally consistent plans;
and (2) miscellaneous guides. References are given below to well-
known series of tourist guides, with specific data on certain recent
volumes. Other important series of tourist guides are: *The American Guide Series* (state guides of the United States, see [459]);
Hanson's *New World Guides to the Latin American Republics;
Blue Guides* [614] (Europe); *Cook's Travellers' Handbooks*
(mostly Europe); *Official Guide to Eastern Asia* [973]; *Guides
Madrolle* [972] (Eastern Asia); guides published by the Touring
Club Italiano (Italy, South America).

Besides tourists guides, reference should be made to guides for special
types of traveler, notably (1) geographers, geologists, and other
scientists (e.g. guides for excursions conducted in connection with

international and other scientific congresses; geological guidebooks
to Western United States published by the United States Geological
Survey); (2) mountaineers (guides usually issued by mountaineering clubs); (3) commercial travelers (e.g. guides published by the
United States Bureau of Foreign and Domestic Commerce). For
many regions not frequented by tourists and not covered by the
standard tourist guides, there are available handbooks of general,
commercial, and practical information for the use of business men,
government officials, and others. References to sources of information regarding some of these are given in Part III.

The American Geographical Society's library contains some 1,500
guidebooks, a representative but by no means exhaustive collection. Modern guidebooks have been neglected by bibliographers.
We know of no comprehensive, world-wide work on the sub-
ject.   [88]

Murray, John, publ. *Handbook[s] for Travellers*. London. 1829–
[1939].   [89]
These guidebooks covering large parts of Europe and Asia were widely
used by British travelers throughout the greater part of the 19th century. The series as a whole has been discontinued with the exception of
the *Handbook for India, Burma and Ceylon* (15th ed., 1939). A vol. on
Japan was issued in 1913.

Baedeker, Karl, publ. *Handbook[s] for Travellers*. Leipzig, 1839–
[1938].   [90]
In the following list of the most recent guides in this famous series *G*
means "in German," all others being available in English (German
guides are referred to only where later than English guides for the
same region, or where English guides do not exist): Austria, with
Budapest, Prague, Karlsbad, and Marienbad (1929); Austria (*G* 1931);
Tirol and Dolomites (1927) (*G* 1929); Belgium and Luxembourg
(1931); Canada, Newfoundland, Alaska (1922); Constantinople, Asia
Minor, Balkan States, Cyprus (*G* 1914); Dalmatia, Western Yugoslavia, Istria, Albania, Corfu (*G* 1929); Egypt and the Sudan (1929);
France: Paris and Environs, 1932 (Suppl. 1937); Northern France,
1909; Southern France, Corsica (1931); Germany (1936); Northern
Germany (1925); Southern Germany (1929); The Rhine (1926); some
15 guides in German for different parts of Germany; Great Britain
(excluding London) (1937); London and Environs (1930); Greece,
Greek Isles, Crete (1909); India, Ceylon, Farther India, Burma, Malay
Peninsula, Siam, Java (*G* 1914); Italy: Northern Italy (1930); Rome

and Central Italy (1930) (*G* 1933); Southern Italy, Sicily, Sardinia, Malta, Corfu (1930); Italy from Alps to Naples (1928) (*G* 1931); Madeira, Canary Islands, Azores (*G* 1934); Mediterranean, Madeira, Morocco, Algeria, Tunisia (1911) (*G* 1934); Norway, Sweden, Denmark, Iceland, Spitzbergen (*G* 1931); Palestine and Syria (1912); Russia, with Teheran and Peking (1914); Spain and Portugal (*G* 1929); Sweden and Finland (*G* 1929); Switzerland (1938); United States, Mexico, Cuba, Puerto Rico, Alaska (1909).

Terry's Guide[s]. Mexico (Hingham, Mass., 1940); Japanese Empire (Boston, Houghton Mifflin, 1933); Cuba (Boston, Houghton Mifflin, 1926).
[91]

MAPS AND CARTOGRAPHY:
BIBLIOGRAPHICAL AIDS

(See also [113–131, 174–186, 194–197].)

This section deals with works covering the whole field of cartography, but more especially with cartobibliographical aids—i.e. aids to the finding of maps and atlases.

Maps are of two kinds: *locational* and *topical* (or special). Locational maps are what we usually have in mind when we say, for example, "map of Maine" or "map of the world," without further qualification. They are locational in the sense that they are most frequently used for locating specific places. Such maps are designed to serve general-utility purposes—to meet the needs of the maximum number of people likely to be concerned with the areas that they cover. Consequently they show a combination of fundamentally important physical and human features, with their names—coast lines, rivers, lakes, cities, political units and their boundaries, sometimes relief, and, depending on the scale, roads, railroads, swamps, etc. (It should be noted that the term "locational map" is also applied to small diagrammatic maps showing the area covered by a larger map (or periodical article) in relation to the surrounding areas.) *Topical* maps, on the other hand, are designed to serve specialized purposes. They show, for example, combinations of all features of especial interest to particular groups of people (as on coast charts) or the distributions of certain specific phenomena only, such as population densities, railways, etc.

Some maps are based directly on surveys; others are compiled from documentary sources. Most of the maps based on surveys are on large scales and are produced by governments. Since maps are the eyes of armies and navies, in many countries the principal agencies that carry on cartographical surveys are connected with the military establishments. The United States and Canada are among the few nations where this has not normally been the case, in peacetime at least.

The majority of maps on scales smaller than 1:250,000 (about 4 miles to the inch) are compiled from other maps, statistics, and other sources.*

The Second World War stimulated an enormous expansion of map-making activities on the part of governments to serve military needs and on that of private organizations to meet the public demand for geographical information pertinent to the war. In the United States the programs of the two principal civilian agencies of the government concerned with the mapping of our coast lines and interior— the United States Coast and Geodetic Survey and the United States Geological Survey, respectively—were stepped up. In addition, widespread ground, marine, and air surveys within but particularly outside the country were undertaken by the Corps of Engineers of the Army, the Army Air Forces, and the Hydrographic Office of the Navy. In order to supply the army with maps a new and extremely enterprising agency, the Army Map Service, was established. It has collected, indexed, and reproduced in various forms an immense quantity of maps of all kinds, including the sets of many of the topographic maps of foreign governments. Compiled maps for military intelligence and for strategic and tactical purposes

* Although discussion of specific maps does not fall within the scope of this volume, mention must be made of one group of maps which, because they are on the uniform scale of 1:1,000,000, for the most part carry layer tints showing altitudes, and conform to a greater or less extent to a standard scheme, are of exceptional utility for general reference and comparative purposes. These so-called "Millionth Maps" are available for most of the land areas of the world outside of northern North America and the U.S.S.R. and also for certain limited areas within the United States, Canada, and the U.S.S.R. They fall into three categories: (1) the official *International Map of the World on the Scale of 1:1,000,000*, the sheets of which are produced by various governments for the territories under their jurisdictions; (2) "Millionth Maps" more particularly of strategic areas produced before and more especially during the Second World War by mapping agencies of various governments; and (3) the American Geographical Society's *Map of Hispanic America on the Scale of 1:1,000,000*, covering the entire American continent and outlying islands south of the United States (107 sheets). Most of the sheets of (1) and (2) have been reprinted by the Army Map Service. See also [94, 547].

were produced in quantity by geographical or cartographical sections in such agencies as the Office of Strategic Services and the intelligence divisions of the Army, Air Forces, and Navy. The State Department and many other governmental agencies also expanded their cartographic work, and, of course, similar developments occurred to a greater or less degree in all of the belligerent states, but particularly in the British Empire.

One result of all this activity has been to render the older guides to the finding of maps somewhat obsolete, since so many new maps have appeared and so many new collections have been built up. Now that the war is over and considerations of security no longer require that much of this work be kept secret, it is to be hoped not only that the new cartographic treasures will be made generally available but also that access to them will be facilitated by the publication of up-to-date cartobibliographical aids.

One should not, however, be so dazzled by the wartime intensification of mapping as to conclude that the aids now available are without value. The accumulation of cartographic information has been going on for centuries and, while the Second World War has increased the pace, it has not brought about anything approaching a complete remapping of the earth's surface. Although far-reaching surveys have been made of many areas of strategic importance and many new "topical" maps have been compiled, a great deal of the feverish map making of the last six years has consisted in reproducing and revamping material that already existed in libraries and had already been dealt with in various aids.

By and large it must be admitted that librarians and bibliographers until recently have been rather stepmotherly in their treatment of maps as compared with their treatment of printed books, periodicals, periodical articles, etc. Atlases, being books, have fared somewhat better. Separate maps are seldom reviewed in the newspapers and popular magazines, and they are not covered satisfactorily for research purposes in most of the general bibliographical works, such as Mudge-Winchell [1], Hirshberg [2], and Besterman [3].

Although maps are not books, the convenient if not wholly appropriate term "cartobibliography" has gained acceptance. Cartobibliographical and related aids are of the following principal kinds:

1) Works which provide general information concerning map collections as units;

2) Cartobibliographies proper, which aim to provide information about maps irrespective of where they are housed or published;

3) Map catalogues, embracing either (a) the maps in a particular collection or group of collections (i.e. union lists), or (b) those issued by a particular governmental agency, commercial publisher, or scientific institution;

4) Index maps to specific maps—i.e., maps which show the areas covered by the individual maps of a particular kind or series, as, for example, the topographic sheets of the United States Geological Survey. These often accompany the map catalogues issued by governmental mapping agencies.

5) Generalized maps (usually compiled from maps of the preceding type) showing the areas covered by all maps of specified kinds or scales without attempting to provide detailed information concerning the individual maps or map sheets themselves.

As the number of works of these various kinds, especially of map catalogues, is large, it would take another volume to deal with them as fully as we deal here with bibliographies and periodicals. A few references are given immediately below to a selection of works which put one on the track of other cartobibliographical aids, and under the appropriate topical and regional headings in Parts II and III a few cartobibliographies of especial note are listed. No attempt has been made to list specific map catalogues and index maps, with one or two exceptions (notably [94]). For these [96–99] should be consulted. It should be noted, however, that most *geographical* aids to research devote attention to maps and, being more selective, are frequently of greater value to the research student in finding the particular maps he wants than are lists and catalogues devoted exclusively to maps themselves. In many of the annotations to the references in this volume we have attempted to indicate something about the treatment of maps, where this is especially worthy of comment. All six of the great current bibliographical repertories devoted to geography [6–11] contain many references to maps.

Geographers and other map users would benefit were someone to compile and publish a classified cartobibliography of *topical* maps. Even if this were confined to the topical maps in readily available world atlases, it would serve an extremely useful purpose. An adjunct to such a work might well be a guide to the finding of maps drawn on different projections. To locate a world or regional map

on a specific projection is often like hunting for a needle in a haystack. [92]
General bibliographical data on maps and cartography: Besterman [3], Vol. 1 (1939), pp. 168–170 (lists various catalogues of collections); for bibliographical data on the science and art of cartography: Wagner [60], 1938, pp. 14–22 (maps); pp. 268–272, 329–330, footnotes on pp. 274–367 (cartography). [93]

Indices of Maps at the Army Map Service Library: Small Scale Series, Scale 1:500,000 and Smaller. (Army Map Service, Corps of Engineers, U.S. Army.) 1st ed. Washington, May, 1945. [94]
Separate index map for each of a large number of map series for all parts of the world.

Platt, R. R. "Official Topographic Maps: A World Index," *Geog. Rev.* [459], 35 (1945), 175–181. [95]
Article descriptive of an accompanying generalized index map of the world in color showing status, as of the outbreak of the present war, of governmental topographic mapping on scales of 1:253,440 and larger.

Olson, E. C., and Agnes Whitmarsh. *Foreign Maps.* New York and London, Harper, 1944. 254 pp. (*Harper's Geoscience Series.*) [96]
Deals mainly with the interpretation of official maps of foreign governments. Contains glossary of geog. terms appearing on maps in many languages; also cartobibl. data: (1) on the principal map series of the United States; (2) on the characteristics of maps of the principal foreign mapping agencies (alph. by nations); (3) a list of refs. (pp. 223–230) covering (a) texts, manuals, and articles, and (b) map catalogues and reports (alph. by nations). As a bibl. aid less detailed but more comprehensive than Thiele [97]. Reviewed in *Geog. Rev.* [476], 34 (1944), 675–676.

Thiele, Walter. *Official Map Publications: A Historical Sketch, and a Bibliographical Handbook of Current Maps and Mapping Services in the United States, Canada, Latin America, France, Great Britain, Germany, and Certain Other Countries.* Chicago, Amer. Library Assn., 1938. 372 planographed pp. [97]
Part I, "A Historical Sketch," covers the history of cartography from ancient times, with especial emphasis on Great Britain and North America. Part II, "Contemporary Government Maps and Mapping Services," contains chapters, with many bibl. data, on the countries indicated in the title, a chapter on Austria, Hungary, Netherlands, and

Norway, and one on the classification, cataloguing, and care of maps. Reviewed in *Geog. Rev.* [476], 29 (1939), 699–701.

Haack, Hermann. "Die Fortschritte der Kartographie (1930–36)," *Geog. Jhrb.* [6], 51 (1936), 230–312; 52 (1937), 3–74.   [98] Topical cl., 6 main hdgs.: (1) genl., incl. periodicals, bibls., international maps, genl. atlases, topographical surveys, etc.; (2) map projections; (3) drafting and reproduction of maps; (4) applied cartography, subcl. according to the various branches of geog.; (5) cartometry; (6) cartography for schools; 1,763 entries; ann.; auth. indexes. See also [1164]. On the progress of topographical surveys Haack gives on pp. 241–248, in the following order refs. to works published in the period 1930–1936 dealing with surveys in the following countries (the number of references is shown in parentheses): Germany (43), Abyssinia (1), Afghanistan (1), Egypt (1), Argentina (1), Belgium (1), China (1), Great Britain (5), British Empire (1), India (1), Canada (1), South Africa (1), Finland (1), France and French colonies (3), Algeria (1), Morocco (1), French Equatorial Africa (1), Madagascar (1), Indo-China (1), Greece (2), Iran (1), Iceland (1), Italy (2), Latvia (1), Netherlands (4), Japan (1), Netherland Indies (1), Scandinavia (1), Norway (1), Austria (4), Peru (2), Poland (2), Rumania (1), U.S.S.R. (1), Sweden (2), Switzerland (3), Spain (1), Yugoslavia (1), Czecho-slovakia (1), Turkey (2), Hungary (3), Uruguay (1), U.S.A. (4). See also [6, Nos. 72–74].

Eckert, Max. *Die Kartenwissenschaft: Forschungen und Grundlagen zu einer Kartographie als Wissenschaft.* Berlin and Leipzig, Verlag Wissensch. Verleger, 1921, 1925. 2 vols.   [99] Monumental work on all aspects of cartography with a wealth of historical and bibl. data. The main hdgs. are: cartography as a science, map projections, surveys, land maps, marine maps.

Carrière, Ludwig. "Unsere Kenntnis der Erde," *Petermanns Mitteilungen* [717], 57, Pt. II (1911), 347–351.   [100] Deals with extent to which topographical and hydrographic mapping had progressed to 1911. Notes on names and scales of large-scale surveys of all countries. Accompanied by a map of the world showing status of cartographical knowledge for both the land and the ocean.

Phillips, P. L. *A List of Works Relating to Cartography.* (The Library of Congress, Division of Maps and Charts.) Washington, 1901. 90 pp. (Reprinted from P. L. Phillips, *A List of Maps of America in the Library of Congress* (The Library of Congress, Division of

Maps and Charts), Washington, 1901, pp. 5–90.) [101]
Alph. by auths., topics, and regions.

Knox, Alexander. *A Guide to Recent Large Scale Maps, Including Both Surveys and Compilations.* (Intelligence Division, War Office.) London, 1899. [102]

Bartholomew, J. G. "The Mapping of the World," *Scottish Geog. Mag.* [626], 6 (1890), 293–305, 575–597; 7 (1891), 124–152, 586–611. [103]
Attempts to indicate in tabular form the best maps available at the time of publ. for all parts of Europe, Africa, Asia, and North America; accompanied by maps of the world and of these four continents showing the extent and quality of existing maps.

Wheeler, G. M. *Facts Concerning the Origin, Organization, Administration, Functions, History, and Progress of the Principal Government Land and Marine Surveys of the World, Being Extracts from the Report on Third International Geographical Congress and Exhibition.* (War Department, Corps of Engineers, U.S. Army.) Washington, 1885. 582 pp. [104]

Periodical

Surveying and Mapping: Journal of the American Congress on Surveying and Mapping
Washington, 1941–. Quarterly after 1942 (previously monthly). Entitled *Bulletin,* 1941–April, 1944. [105]

Map collections in the United States: L. C. Karpinski, "Cartographical Collections in America," *Imago Mundi* [177], 1 (1935), 62–64 (short list of collections of interest to the student of old maps).— *Map Collections in the District of Columbia,* prepared by the [U.S.] Geological Survey for the Federal Board of Surveys and Maps, revised to Sept., 1938, 56 mimeo. pp.—*Geog. Rev.* [476], 31 (1941), 270–271.—See also [38].
Though not the largest map collection in the country, that of the American Geographical Society is critically selected and conveniently organized for purposes of geographical research. It contains some 130,000 maps. Both separate sheets and those contained in books are covered by a special map card catalogue (alph. by areas and topics) and cards for maps are entered in the Research Cata-

logue [20]. The Society also has a notable collection of carto-bibliographical aids. [106]

Boggs, S. W., and Dorothy C. Lewis. *Classification and Cataloging of Maps and Atlases.* New York, Special Libraries Assn., 1945. 175 pp. [107]
Bibl. on cataloguing and classification of maps: pp. 166–173.

BIBLIOGRAPHIES OF ATLASES

Mudge-Winchell [1], a 102–103, 110, 333–338; b 47–48; c 67; d 61–62.—Haack [98], 1936, pp. 264–265, 269–270, 297–300.—On six representative European atlases: note in *Geog. Rev.* [476], 27 (1937), 161–163 (deals with Stieler [123], atlases of France [706], and Czechoslovakia [800], *Allt för Atlas Världsatlas* [124], and 2 German regional atlases). [108]

"World Reference Atlases: A Survey of Current Resources," 8 pp. *Bull. Amer. Soc. for Professional Geographers* [485], Vol. 3 (1945), Nos. 5–6. [109]
Prepared by a committee of the American Society for Professional Geographers, this paper contains a discussion of the "essential qualities for world atlases," a review of selected world atlases, and a brief bibl. of works concerning atlases.

Ristow, W. W. "A Survey of Recent Atlases," *Library Jour.,* 70 (1945), 54–57, 100–103. [110]

Joerg, W. L. G. "Post War Atlases," *Geog. Rev.* [476], 13 (1923), 583–598. [111]
Deals more especially with *Times* [119], *Daily Telegraph* (1922?), Harmsworth's (1921?), Vivien de St. Martin and Schrader's [120], Westermann's (1922), and Italian Touring Club [122] atlases.

Phillips, P. L. *A List of Geographical Atlases in the Library of Congress.* 4 vols. Washington, 1909–1920. [112]
The only comprehensive bibl. of atlases. Lists 5,324 atlases dating from the Middle Ages through 1919. Divided into (1) basic chron. lists of atlases; (2) alph. auth. lists; (3) alph. indexes. For many but not all of the atlases there are analytical notes listing individual maps: primarily "maps relating to America, plans of cities throughout the world, and material of specific interest not usually found in atlases." The indexes

cover only the maps listed in the analytical notes, which are less complete for modern than for early atlases. In the indexes topical subhdgs. are given under areal hdgs., there being no independent topical hdgs.

GENERAL WORLD ATLASES

Atlases are of four kinds:

1) *General world atlases,* designed to meet widely diverse public needs. These are discussed in this section.
2) *Topical* or *special world atlases,* which deal with particular subjects (e.g. history, meteorology) and are listed under the appropriate topical headings in Part II.
3) *General regional* and (4) *topical regional* atlases, of which the scope is limited to specific regions. These are listed under the appropriate regional headings in Part III.

In choosing an atlas as an aid to geographical research one should consider its size, the relative number of locational and topical maps (see above, p. 83) that it contains, the scales of and areas covered by the maps of both types, and the subjects illustrated by the topical maps. No less important are the accuracy and clarity of the individual maps, the balance of the work as a whole, and the nature of the text, indexes, and other noncartographical adjuncts. Criteria for judging the quality and utility of general atlases, particularly from the reference librarian's point of view, are indicated in Mudge-Winchell [1], a 332–333; Ristow [110], 1945, and "World Reference Atlases" [109], 1945.

The following references are to a selection of 18 general world atlases in frequent use at the American Geographical Society. While other atlases, equally good in many particulars, might also be cited, it is believed that this selection includes the most outstanding works. The following table presents some salient facts about the 18 atlases for purposes of rapid comparison. It shows in the first column the sizes of the atlases in terms of space devoted to maps—a fairly satisfactory indication of the amount of cartographic information that an atlas provides. The second and third columns show the map space devoted to locational maps, and the fourth and fifth columns, respectively, show the percentages of the total map space devoted to topical maps and to maps of the home country. Map space was calculated by multiplying the number of pages devoted to maps by the area of the page in square centimeters. The data are based on an examination of the latest editions in the collection of the American Geographical Society.

SELECTED WORLD ATLASES

Atlas	Serial Number	MAP SPACE SQUARE METERS Total	Loca-tional Maps	PERCENT OF TOTAL MAP SPACE Loca-tional Maps	Topi-cal Maps	Maps of Home Country	Home Countr
		1	2	3	4	5	6
1. *New World*	[115]	47	35	74	26	38	U.S.A.
2. Soviet	[121]	47	12	26	74	72	U.S.S.R.
3. Rand McNally	[114]	44	28	63	37	74	U.S.A.
4. Andree	[125]	35	31	88	12	19	German₹
5. *Times*	[119]	33	30	91	9	11	British I
6. *Atlas Internazionale della Consociazione Turistica* (Italian)	[122]	27	26	95	5	7	Italy
7. *Universel*	[120]	25.2	24.9	99	1	11	France
8. Stieler	[123]	23	23	100	0	7	German
9. Vidal-Lablache	[126]	20	9	46	54	15	France
10. Philip [b]	[118]	18	16	89	11	14	British
11. *Citizen's* [c]	[117]	17	14	87	13	19	British
12a. *Encyclopædia Britannica*, 1945	[116]	17	14	82	18	36	U.S.A.
b. *Encyclopædia Britannica*, 1942	[116]	14	14	100	0	44	U.S.A.
13. Meyer	[128]	14	13	92	8	16	German
14. Goode	[129]	13	7	58	42	19	U.S.A.
15. *Världsatlas*	[124]	9	7	74	26	26	Scandin
16. *Oxford Advanced*	[130]	9	5	59	41	8	British
17. Sydow-Wagner	[131]	6.9	4	58	42	18	German
18. *Bartholomew's Handy Reference*	[127]	6.7	6.4	90	10	19	British

[a] Maps covering the whole of "Mitteleuropa" are counted among maps of Germany.
[b] Data based on 1931 edition.
[c] Data based on 1935 edition.

In many atlases certain maps designated as "political" and "physical" were counted as locational rather than as topical. Such maps are not topical in the sense in which maps showing the distribution of rain-fall or of votes in an election would be "physical" or "political." The conventional "political" map in an atlas in reality is a locational map on which political units are more emphasized than are relief, drainage, etc., and the reverse is true of the conventional "physical" map. The primary purpose of maps of both types, however, is to enable the reader to locate specifically named places.

In many atlases separate "political" or "physical" maps, or both, are provided for certain areas (frequently the world and the continents) in addition to locational maps not designated as either "political" or "physical." Where this is the case, the latter are called "basic locational" maps (abbreviated "b.l.") in the annotations below.

In the table the atlases are listed in decreasing order of size as measured in terms of map space. It is also of interest to rank them as follows, according to percentage of map space devoted to (1) *locational maps:* Stieler (100), *Encyclopædia Britannica, 1942* (100), *Universel* (99), Italian (95), Meyer (92), *Times* (91), *Handy Reference* (90), Philip (89), Andree (88), *Citizen's* (87), *Encyclopædia Britannica, 1945* (82), *New World* (74), *Världsatlas* (74), Rand McNally (63), *Oxford Advanced* (59), Goode (58), Sydow-Wagner (58), Vidal-Lablache (46), Soviet (26); (2) *maps of the home country:* Stieler (7), *Oxford Advanced* (8), *Times* (11), *Universel* (11), Philip (14), Vidal Lablache (15), Meyer (16), Sydow-Wagner (18), Andree (19), *Citizen's* (19), *Handy Reference* (19), *Världsatlas* (26), *Encyclopædia Britannica, 1945* (36), *New World* (38), *Encyclopædia Britannica, 1942* (44), Soviet (72), Rand McNally (74).

Some facts of practical value to the geographer that are brought out by these figures and in the annotations to the atlases listed below may be briefly stated. Among the larger atlases (20 or more square meters of map space), the most generally useful are Andree, *Times*, *Universel*, Italian, and Stieler. In none of these is there disproportionate (over 20 percent) emphasis on the home country. The first two, but more especially Andree, have the advantage of a goodly number of well-selected topical maps. The others are predominantly locational. The two American atlases (*New World* and Rand McNally) are of value primarily for their detailed locational maps of the several states of the United States. The topical maps and the locational maps of foreign countries in these atlases are less distinguished. The Soviet atlas is unique by reason of its wide variety of topical world maps and its locational and topical maps of the

U.S.S.R. It contains, however, no regional maps of other parts of the world. Vidal-Lablache is partly historical; the topical maps pertaining to modern geography are varied and excellent.

Among the smaller atlases (less than 20 square meters) Goode, Sydow-Wagner, and the *Oxford Advanced* are school atlases, especially useful for their topical maps, which are well distributed both regionally and in terms of geographical subjects. Actually more map space is devoted to topical maps in Goode (5.0 sq. m.) than in Andree (4.2), and in both Sydow-Wagner (3.9) and *Oxford Advanced* (2.9) than in *Times* (3.0). An excellent working combination would be Andree and Goode, the former for its clear, comprehensive hachured locational maps and its topical maps more particularly for Europe, and the latter for its layer-tinted locational maps and its topical maps more particularly for the United States. From the point of view of beauty of workmanship and of geographical originality displayed in its topical maps, *Världsatlas* stands in a class by itself. The other smaller atlases are all primarily locational. Except for Meyer, the quality of their maps does not approach that of *Times,* Andree, Stieler, or the Italian atlas. The shortcomings of the locational maps in *Encyclopædia Britannica Atlas* (a large atlas in terms of bulk but not of map space) are outbalanced by the utility of the statistical tables in the 1942 edition and of the topical world maps in the 1945 edition. Meyer and *Handy Reference* are serviceable for quick reference on the reader's desk or in his lap in cases where a larger atlas is not required.   [113]

LARGE ATLASES

Λmerican and British

Rand McNally Commercial Atlas and Marketing Guide. 76th ed. New York and Chicago, Rand McNally, 1945. 568 pp. (also a Road Atlas of United States, Canada, and Mexico in pocket). 39 x 53 cm.   [114]
In two parts: (1) The United States and Its Possessions; (2) Foreign Countries.

Map space (excluding Road Atlas), 44 sq. meters; U.S.A., 72 percent; locational maps, 63 percent (b.l. maps of foreign countries carry hachures or no relief; b.l. maps of U.S. states tinted according to density of population, no relief); topical maps, 37 percent (mostly U.S.A.; especially data designed to meet needs of business men, e.g. transportation routes, economic data, etc.). Separate index for each b.l. map of U.S. states, giving populations, altitudes, and other facts of commercial interest (e.g. railroad connections, post offices, banking towns, etc.) as well as locations; genl. index to cities for b.l. maps of foreign countries, giving populations.

The New World Loose-Leaf Atlas. 6th ed. New York, Hammond. New plates currently issued for loose-leaf binder. *Ca.* 416 pp. (also Postal Guide, at back). 34 x 51 cm. [115]
Map space, 47 sq. meters; U.S.A., 38 percent; locational maps, 74 percent (b.l. maps, some with hachures, mostly without relief; physical maps of continents and states in U.S., layer tinted); topical maps, 26 percent (world, continents; climate, econ. geog., population, history, etc.). Separate indexes for each b.l. map (for U.S. states, separate indexes to counties and incorporated places, giving populations, and to smaller places giving location only); no genl. index.

Encyclopædia Britannica World Atlas. . . . Prepared by G. D. Hudson under the editorial direction of Walter Yust. Chicago, etc., Encyclopædia Britannica, 1942. 112 pp. of maps. 139 pp. of index, etc. 269 pp. of "Geographical Summaries," etc. 31 x 41 cm. (Also in an edition published by C. S. Hammond, New York, 1945, with 28 additional pp., but without the Geographical Summaries.) [116]
Map space (excluding maps in "Geographical Summaries"), 16 sq. meters; U.S.A. 44 percent; locational maps, 100 percent (b.l. maps, no relief; layer-tinted physical maps of world, continents, and U.S.). Index gives populations as well as locations. Reviewed by M. J. Proudfoot in *Geog. Rev.* [476], 33 (1943), 347–348. On "Geographical Summaries" see [50]. The 1945 edition contains 25 pp. of world maps (physical geog., population, languages, economic geog.).

The Citizen's Atlas of the World. J. G. Bartholomew, ed. 5th ed. Edinburgh, The Geographical Institute, 1935. 192 pl. 103 pp. 24 x 36 cm. An 8th ed. was published in 1944. [117]
Map space, 17 sq. meters; British Isles, 19 percent; locational maps, 87 percent (gray hachures or no relief); topical maps, 13 percent (world, Europe, British Isles; climate, ocean currents, time zones, communications, etc.). Text incl. etymology of place names, principal journeys of exploration, etc. Index.

Philip's International Atlas. . . . George Philip, ed. London, Philip, 1931. 158 pl. 98 pp. 28 x 41 cm. An "Interim Edition" was published in 1944. [118]
Map space, 18 sq. meters; British Isles, 14 percent; locational maps,

89 percent (gray hachures); topical maps, 11 percent (world, Europe, British Isles). No text. Index.

The Times Survey Atlas of the World: A Comprehensive Series of New and Authentic Maps Reduced from the National Surveys of the World and the Special Surveys of Travellers and Explorers with General Index of Over Two Hundred Thousand Names. Prepared under the direction of J. G. Bartholomew. London, *The Times,* 1922. 112 pl. 32 x 46 cm.   [119]
Map space, 33 sq. meters; British Isles, 11 percent; locational maps, 91 percent (b.l. maps, layer tinted; political maps of continents); topical maps, 9 percent (world, continents, British Isles; ethnography, population, railways, industries). No text. Index (in separate pamphlet).

FOREIGN

Vivien de Saint-Martin and Schrader. *Atlas universel de géographie.* New ed. by F. Schrader. Paris [1943?]. 80 pl. 88 pp. 35 x 45 cm.   [120]
Map space, 25 sq. meters; France, 11 percent; locational maps, 99 percent (b.l. maps, brown hachures; layer-tinted physical maps of world and continents). Index.

Bol' shoĭ sovetskiĭ atlas mĭra [Great Soviet World Atlas]. Moscow [U.S.S.R. government publication], 1937, 1939. 2 vols.   [121]
Vol. 1, ed. under the direction of V. E. Motylev; Part 1, World Maps (pls. 1–83); Part 2, Maps of the U.S.S.R. (pls. 84–168). Vol. 2, ed. under the direction of S. A. Kutaf'ev; Part 1, Political-Administrative, Survey, Economic, and Physical Maps of U.S.S.R. (pls. 9–134); Part 2, Historical Maps of the Civil War in the U.S.S.R. (pls. 135–143). A third vol. for other parts of the world was planned but not published. Map space, 47 sq. meters; U.S.S.R., 72 percent; locational maps (mostly the "survey maps" of U.S.S.R. in Vol. 2), 26 percent (layer tints); topical maps, 74 percent (world and U.S.S.R. only; these are of great variety and ingenuity, covering physical and human geog., with special emphasis on natural resources, manufacturing, agriculture, and other economic circumstances). Reviewed by G. B. Cressey in *Geog. Rev.* [476] 28 (1938), 527–528.
   For Vol. 1, an English transl. of the titles of and legends on the maps is available (*Great Soviet World Atlas,* Volume 1, transl. under the direction of G. B. Cressey, Syracuse Univ., Syracuse, N.Y., 1940; planigraphed). Vol. 2, of which there are only two or three copies of the original in the United States, has been reproduced in color photog-

raphy in a limited edition with English transl. of the table of contents, by the United States Office of Strategic Services.

Atlante internazionale della Consociazione Turistica Italiana. Ed. under the direction of L. V. Bertarelli, O. Marinelli, and P. Corbellini. (Touring Club of Italy.) 5th ed. Milan, 1938. 95 pl. 248 pp. 32 x 49 cm. [122]
Map space, 27 sq. meters; Italy, 7 percent; locational maps, 95 percent (b.l. maps, brown hachures; political and layer-tinted physical maps of world and continents); topical maps, 5 percent (incl. communications, ethnography of Europe). Index. Sources of information indicated on backs of plates. Comparable to Stieler [123], except that plates are larger and less crowded. 4th ed. reviewed in *Geog. Rev.* [476], 24 (1934), 162–163.

Stieler: Grand Atlas de géographie moderne. 10th ed.; édition internationale. Gotha, Justus Perthes, 1934–. 114 pl. 25 x 40 cm. [123]
Map space, 23 sq. meters; Germany and Central Europe, 7 percent; locational maps, 100 percent (brown hachures; extremely detailed). Index.
The German edition is known as *Stielers Handatlas.* In the International Edition, which was still incomplete in 1939, the maps are in the languages of the countries mapped.

Världsatlas. . . . Ed. under the direction of Axel Elvin. Stockholm, Åhlén & Åckerlund, 1934. 50 pl. 248 pp. 23 x 39 cm. (Revised ed. of *Nordisk världsatlas,* 1926.) [124]
Map space, 9 sq. meters; Scandinavia, 26 percent; locational maps, 74 percent (b.l. maps, land areas buff tinted, political boundaries in color, purple hachures; layer-tinted physical maps); topical maps, 26 percent (world on equal area projections, Europe, U.S.A.; physical and human geog., population distribution, etc.). Comprehensive text. Index.

Andrees allgemeine Handatlas. 8th ed. Ed. by Ernst Ambrosius. Bielefeld and Leipzig, Velhagen & Klasing, 1930, 224 pl. 29 x 45 cm. [125]
Map space, 35 sq. meters; Germany and Central Europe, 19 percent; locational maps, 88 percent (brown hachures; somewhat less detailed than Stieler); topical maps, 12 percent (world, Europe, Germany; physical and economic geog., populations, religions, etc.) No text. Index in separate vol. (*Namenverzeichnis zu Andrees Handatlas,* 643 pp.).
A rev. ed. in 100 plates, from which the larger locational maps were omitted, was published in 1937 (index in same vol.).

Histoire et géographie: Atlas général Vidal-Lablache. New ed. Paris, Colin, 1925. 131 pl. 48 pp. 28 x 38 cm. [126]
History (pls. 1–52); geog. (pls. 52a–131). Map space, 20 sq. meters; France, 15 percent; locational maps, 46 percent (brown hachures); topical maps, 54 percent (majority historical; also physical, human, and economic geog.). Index.

SMALLER ATLASES

The Handy Reference Atlas of the World. John Bartholomew, ed. 14th ed. (Geographical Institute.) Edinburgh and London, 1940. 234 pl. 220 pp. 15 x 21 cm. [127]
Map space, 6.7 sq. meters; British Isles, 19 percent; locational maps, 90 percent (mostly without relief; a few with gray hachures); topical maps, 10 percent. Index.

Meyers geographischer Handatlas. 7th ed. Leipzig, Bibliographisches Institut, 1928. 92 pl. 200 pp. 15 x 24 cm. [128]
Map space, 14 sq. meters; Germany and Central Europe, 16 percent; locational maps, 92 percent (b.l. maps, brown hachures; for continents, political and layer-tinted physical maps); topical maps, 8 percent (world, Europe, Germany). Index.

SCHOOL AND COLLEGE ATLASES

Goode, J. P. *Goode's School Atlas: Physical, Political, and Economic.* New York, Chicago, Rand McNally, 1943 (latest ed.). 286 pp. (173 pp. of maps). 25 x 29 cm. [129]
Map space, 13 sq. meters; U.S.A., 19 percent; locational maps, 58 percent (layer tinted; many city plans on small scales); topical maps, 42 percent (especially for world, U.S., and Canada; cover wide range of subjects, incl. map projections, physical geog., soils, vegetation, agriculture, industries, etc.). Index.

Atlas of World Maps for the Study of Geography in the Army Specialized Training Program. Headquarters, Army Service Forces, 1943. (*Army Service Forces Manual* M-101.) 30 pl. 13 pp. 55 x 26 cm. [129a]
Map space, 4.3 sq. meters. Exclusively world maps covering population, languages, religions, and various aspects of physical, economic, and political geog. See also [314].

General Aids 99

The Oxford Advanced Atlas. John Bartholomew, ed. 6th ed. London, etc., Oxford Univ. Press, 1940. 96 pl. 36 pp. 25 x 37 cm.   [130]
Map space, 9 sq. meters; British Isles, 8 percent; locational maps, 59 percent (layer tints); topical maps, 41 percent (world, continents, especially Europe, British Isles; climate, population, occupations, economic geog., etc.). No text. Index.

Sydow-Wagners methodischer Schul-Atlas. 20th ed. Ed. by H. Haack and H. Lautensach. Gotha, Justus Perthes, 1932. 62 pl. 15 pp. 18 x 31 cm.   [131]
Map space, 6.7 sq. meters; Germany and Central Europe, 18 percent, locational maps, 58 percent (layer tints and brown hachures); topical maps, 42 percent (well distributed regionally but with main emphasis on Europe and Germany; cover a wide range of geog. subjects). List of sources in text. Index (separate pamphlet).

II. TOPICAL AIDS

Though none of the works listed in this section is restricted in scope to any specified region or regions, each one is pertinent to the study of a particular aspect of geography rather than to the field as a whole. Some are general aids covering the sciences, social studies, and humanities in the large. Others are more specialized aids to work in subjects closely allied to geography, and still others deal with specific branches of geography as such. In general, no attempt is made to include periodicals dealing with the various fields related to geography, as their number is legion and data concerning them may readily be found in the bibliographies to which reference is made.

Data on maps pertaining to special branches of geography will be found in Haack [98], 1936–1937; [1164], 1942.   [132]

HISTORICO-GEOGRAPHICAL STUDIES

Geography and history merge together over broad realms. For guidance to geographical studies bearing on the ages before the dawn of recorded history, aids to geological, anthropological, and archeological research should be consulted. Here, by historico-geographical studies, we mean investigations based primarily on written documents and maps. These fall into three main categories:

1) Studies of human history in the light of geographical circumstances, as, for example, studies that seek to narrate or interpret the unfolding of past events in terms of influences of the natural environment. These might be called studies in *geographical history*. References to a selection of important works by geographers and historians bearing on this subject will be found in H. E. Barnes, *The New History and the Social Studies* (New York, Century, 1925), chap. 2, "The Relation of Geography to the Writing and Interpretation of History," pp. 40–75.

2) Studies which seek to reconstruct the geographical circumstances of the past, either as they existed at a particular time or as they developed during longer or shorter periods. This field is usually called *historical geography*. See [6, Nos. 20–25].

3) Studies of the past development of geographical knowledge, customarily known as the *history of geography.*

(Logically a fourth category would comprise the "geography of history" (or of historiography), which might deal with the geographical distribution of historical investigations and with the influences of geographical circumstances on the growth of historical thought.) The following references are classified under three main headings: (1) aids of general utility in connection with historico-geographical studies, and aids of value primarily in connection with (2) historical geography, and (3) the history of geography. Since, however, these three subjects are closely interrelated, it is recommended that the student of any one of them consult not only the references for that subject but those given under the other headings as well.

[133]

General Historico-Geographical Aids

History: Mudge-Winchell [1], ancient and prehistory, a 345–347, b 49, c 68; annuals, a 343–344; bibliographies, general, a 341–342; bibliographies, Europe, d 63; bibliographies, medieval, a 342; bibliographies, World War I, a 343, World War II, d 63–64; dictionaries, general, a 340, c 67–68, d 63; dictionaries, ancient geography, a 323; lists, outlines, tables, a 344, c 67–68; national, by countries, a 348–363, b 49–52, c 68–72, d 64–69.

[134]

Davis, E. J., and E. G. R. Taylor. *Guide to Periodicals and Bibliographies Dealing with Geography, Archaeology, and History.* London, G. Bell, 1938. 22 pp. (*Historical Assn. Pamphlet*, No. 110.

[135]

Deals with sources of information (for the most part English publs.) regarding current work in geog., archaeology, and history, "with a view to facilitating study in the zones of overlap between the three subjects."

Dutcher, G. M., and others, eds. *A Guide to Historical Literature.* New York, Macmillan, 1937. 1,250 pp. [136]

26 sections, each dealing with a topic (e.g. hist. and auxiliary sciences, genl. hist., exploration and colonial expansion), a chron. period (e.g. contemporary times, 1871–1930), a region during a chron. period (e.g. Near East in ancient times), or a region (e.g. France, Africa); each section subcl. consistently, "as far as varying conditions have permitted,"

by subjs. (incl. bibl.; encyclopedias; geog. and atlases; ethnography; source books, collections of sources; genl. hists.; hists. of special periods, regions, or topics; constitutional, economic, social, and cultural hist.; biography; govt. publs.; academy publs.; periodicals); fully ann. by a large group of contributors; index to auths., periodicals, and academy publs.

Coulter, Edith M., and Melanie Gerstenfeld. *Historical Bibliographies: A Systematic and Annotated Guide.* Berkeley, Calif., Univ. of California Press, 1935. 218 pp.   [137]
Cl. in the main by continents, subcl. by countries (United States by states); further subcl. by topics; also sections on genl. bibls., ancient world, voyages and travels, colonial possessions, and, under "Europe," on medieval, modern, and religious hist.; 775 entries; ann.; genl. index.

Historical Geography

(See also [6, Nos. 20–25].)

On historical geography at recent international congresses: note in *Geog. Rev.* [476], 24 (1934), 504–505.—On gazetteers of ancient geography: Mudge-Winchell [1], a 323.   [138]

Pauly, A. F. von. *Pauly's Real-Encyclopädie der classischen Altertumswissenschaft.* New ed. begun by Georg Wissowa, ed. by Wilhelm Kroll and Karl Mittelhaus. Stuttgart, Metzler, 1894–[1939]. 42 vols. to date in 2 series (1, A-Ph; 2, Ra-Tu), and 6 Supplements.   [139]
An immense wealth of detailed bibl. information bearing on ancient geog. will be found in the articles in this monumental encyclopedia dealing with countries, provinces, cities, etc. See also [154] and [175].

HISTORICAL ATLASES

Treharne, R. F. *Bibliography of Historical Atlases and Hand-Maps for Use in Schools.* London, G. Bell, 1939. 24 pp. (*Historical Assn. Pamphlet,* No. 114.)   [140]
Deals with foreign as well as with British and American atlases. 3 main parts: (1) classroom atlases, (2) hand maps for class use, (3) reference atlases and hand maps; 104 entries; ann.

Spruner von Merz, Karl, and others. [Historical Atlas.] Gotha, Justus Perthes, 1846–[1938].   [141]
The first ed. of this great atlas, entitled *Dr. Karl von Spruners historisch-*

geographischer Hand-atlas, was issued in 3 parts: (1) *Atlas antiquus,* 1850; (2) *Historisch-geographischer Hand-atlas zur Geschichte der Staaten Europas vom Anfang des Mittelalters bis auf die neueste Zeit,* 1846; (3) *Historisch-geographischer Hand-atlas zur Geschichte Asiens, Afrikas, Amerikas und Australiens,* 1851. A second ed. of the whole work appeared in 1854–1855. Third eds. of Parts 1 and 2, ed. by Theodore Menke, were publ. in 1865 and 1880, respectively, the latter under the title *Spruner-Menke Hand-atlas für die Geschichte des Mittelalters und der neueren Zeit* (90 colored maps, 376 inset maps, 42 pp. text; 27 x 40 cm.). A new ed. of Part 1, entitled *Atlas antiquus: Atlas zur Geschichte des Altertums,* ed. by Wilhelm Sieglin, was begun in 1893 and continued by Max Kiessling; to be completed in 8 fascicules.

Shepherd, W. R. *Historical Atlas.* 7th ed. New York, Holt, 1929. 216 pp. colored pls. (*ca.* 250 maps), 127 pp. contents, index, etc. 18 x 26 cm.
[142]
Based in the main on German historical atlases, the maps were printed in Germany. Emphasis largely on European political and military hist., though America and other parts of the world as well as economic and cultural hist. are also represented.

Putnam's Historical Atlas: Medieval and Modern . . . , by Ramsay Muir, George Philip, and R. M. McElroy. 6th ed. New York and London, Putnam, 1927. 96 pls. with 229 colored maps and diagrams; 132 pp. text, contents, index, etc. 24 x 29 cm.
[143]
Contains a section on the economic hist. of the United States illustrated by uncolored maps, also an introduction illustrated by 40 uncolored maps and plans, dealing with the development of Europe and the United States. This atlas is distinguished by its emphasis upon (1) the physical basis of hist., (2) economic and other aspects of hist. as well as the political aspect, and (3) the expansion of European influence or control throughout the non-European world (preface).

Schrader, F. *Atlas de géographie historique.* New ed. Paris, Hachette, 1911. 55 pls. in color on backs of which is historical text containing many detailed figures, diagrams, etc. 26–38 cm.
[144]
Emphasis mainly on ancient, medieval, and modern European hist.; also plates devoted to maps of the world at various dates, Asia, America, colonial expansion, etc.

Droysen, G. *Allgemeiner historischer Handatlas.* . . . Bielefeld and Leipzig, Velhagen u. Klasing, 1886. 88 colored pl.; 94 pp. text, contents. 29 x 42 cm.
[145]
Emphasis on Europe and Germany.

History of Geography

(See also [6, Nos. 26, 27].)

The history of geography seeks to record how geographical data have been acquired through exploration and field studies, how they have been recorded and interpreted for scientific, educational, and popular purposes, and how geographical knowledge has been disseminated. Geographers are naturally interested in the history of geography as providing background and perspective for a better understanding of the growth and nature of their own science. The history of geography, however, embraces an immense complex of activities and ideas, which have played an extremely important part in the larger history of civilization as a whole, and hence is a field that has been cultivated fully as intensively by historians, classicists, Orientalists, etc., as by geographers as such.   [145a]

GENERAL WORKS ON THE HISTORY OF SCIENCE AND OF GEOGRAPHY

History of science: Mudge-Winchell [1], a 168, d 32.   [146]

Sarton, George. *Introduction to the History of Science.* Baltimore, Williams & Wilkins. Vol. 1, 1927, 850 pp. Vol. 2 (in 2 separately bound parts with continuous pagination), 1931, 1,302 pp. (*Carnegie Instn. of Washington Publ.,* No. 376.)   [147]
This compendium, containing a wealth of bibl. refs. both to the original sources and to secondary works, covers the development of science throughout the world from Homeric times to A.D. 1300. The primary divisions of the text are by chron. periods, subcl. by civilizations (e.g. Greeks, Arabs, Europeans, Chinese) and by topics, incl. geog.; genl. index in each vol.

Josephson, A. G. S. *A List of Books on the History of Science.* . . . (The John Crerar Library.) Chicago, 1911. 297 pp.   [148]
5 main parts: (1) genl. works, (2) social sciences, (3) physical sciences, (4) natural sciences (incl. geog.), (4) medical sciences, each subcl. by topics; for many works contents analyzed; genl. index.

Periodical

Isis: An International Review Devoted to the History of Science: Founded and Edited by George Sarton. Official Quarterly Journal of the History of Science Society

Cambridge, Mass. (formerly Brussels, Bruges), 1913-. (Suspended July, 1914—Aug., 1919.) Subtitle varies. [149]
Contains comprehensive critical bibls.

Broad survey of the literature on the history of geography: Wagner [60], 1938, pp. 1–5.—See also [1166]. [150]

Dickinson, R. E., and O. J. R. Howarth. *The Making of Geography.* Oxford, Clarendon Press, 1933. 264 pp. [151]
Selected list of books and articles which deal specifically with the hist. of geog. on pp. 254–258. Reviewed in *Geog. Rev.* [476], 23 (1933), 701–702.

Günther, Siegmund. *Geschichte der Erdkunde.* Leipzig and Vienna, Deuticke, 1904. 354 pp. (*Die Erdkunde,* ed. by Maximilian Klar, Pt. 1.) [152]
An abundance of bibl. refs. in the footnotes for period prior to 1800.

HISTORY OF GEOGRAPHY FOR THE MOST PART PRIOR TO THE NINETEENTH CENTURY

A section devoted to the ancient geography of India appeared regularly in each volume of *Annual Bibliography of Indian Archæology* (covering 1926–1937), Leyden; see also list of "important treatises and writings on ancient Indian geography," by B. C. Law in the volume for 1935, pp. 12–20.—For bibliographical data on English geographical literature to 1650, see E. G. R. Taylor's *Tudor Geography, 1485–1583* (London, Methuen, 1930; 299 pp.) and *Late Tudor and Early Stuart Geography, 1583–1650* (London, Methuen, 1934; 333 pp.). [153]

Gisinger, F. "Geographie," in Pauly [139], Supplementband IV, 1924, cols. 521–685. [154]
Detailed survey of ancient and medieval geog. with wealth of bibl. refs.

Kramers, J. H. "Djughrāfiyā" [Geography], in Houtsma, *Encyclopaedia of Islam* [922], Supplement, 1938, pp. 61–73. [155]
Article on the history of Mohammedan geog. literature with many refs. in text and a short bibl. of outstanding editions and translations of the sources, bibls., and other secondary works.

Taeschner, Franz. *Die geographische Literatur der Osmanen.* Leipzig, Brockhaus, 1923. (Separate from *Zeitschrift der Deutschen Morgenländischen Gesellschaft,* Vol. 2, No. 1.) [156]

Bibl. refs. in footnotes. Covers period since 1800 as well as earlier literature.

Kretschmer, Konrad. "Die Literatur zur Geschichte der Erdkunde vom Mittelalter an (1907–25), *Geog. Jhrb.* [6], 41 (1926), 122–189. [157]
Deals almost entirely with studies of the history of geog. prior to 1800. 6 parts: (1) genl., (2) early Christian Middle Ages, (3) Arabs and other Orientals, (4) scholastic period, (5) age of discovery (subhdgs.: explorations, cartography), (6) age of measurements; 708 entries; ann.; auth. index.

Wright, J. K. *The Geographical Lore of the Time of the Crusades: A Study in the History of Medieval Science and Tradition in Western Europe.* New York, 1925. 584 pp. (*Amer. Geog. Soc. Research Ser.,* No. 15.) [158]
Deals primarily with 12th and 13th centuries, with introductory chapters on earlier periods. Bibl. (alph. by auths.) on pp. 503–543, preceded by a bibl. note (pp. 491–502) designed to give a rapid introduction to the outstanding publs. in the field of ancient and medieval geog. to A.D. 1250 (topical cl.).

Atkinson, Geoffroy. *La Littérature géographique française de la Renaissance: répertoire bibliographique. . . . Description de 524 impressions d'ouvrages publiés en français avant 1610, et traitant des pays et des peuples non européens, que l'on trouve dans les principales bibliothèques de France et de l'Europe occidentale.* Paris, Picard, 1927. 565 pp. *Supplément,* 1936, 88 pp. [159]
Main list chron. by titles; 7 alph. indexes incl. index to geog. names.

HISTORY OF GEOGRAPHY FOR THE MOST PART SINCE 1800

Shortly after the First World War three reviews of the progress of geographical work in Europe were published by the American Geographical Society: W. L. G. Joerg, "Recent Geographical Work in Europe," *Geog. Rev.* [476], 12 (1922), 431–484 (also printed separately); Sir John Scott Keltie, *The Position of Geography in British Universities,* 1921 (*Amer. Geog. Soc. Research Ser.* No. 4); and Emmanuel de Martonne, *Geography in France,* 1924 (*ibid.,* No. 4a). With these may also be mentioned Roberto Almagrià, *La*

Geografia, 2d ed., Rome, 1922 (*Guide Bibliografiche*). All of these contain bibliographical references.—See also [950, 974, 977]. [160] On geographical work in certain European countries during the Second World War articles or notes have appeared in the *Geographical Review* [476], 36 (1946), as follows: Belgium, pp. 155–156; France (by Jean Gottmann), pp. 80–91; Germany (by Eric Fischer), pp. 92–100, (by T. R. Smith and L. D. Black), pp. 398–408; Great Britain (by L. S. Wilson), pp. 597–612; Norway, pp. 326–327; U.S.S.R. (by Jean Gottmann), pp. 161–163. [161]

Hartshorne, Richard. *The Nature of Geography: A Critical Survey of Current Thought in the Light of the Past.* Lancaster, Penn., Assn. of Amer. Geogs., 1939. 488 pp. Also in *Annals Assn. of Amer. Geogs.* [479], 29 (1939), 171–658. [162]
Scholarly and comprehensive survey and critique of writings in the field of geog. bearing particularly on methodology. List of refs. to works cited in the text, pp. 178–197: 2 main parts: (1) hist. of geog. thought prior to 1900 (genl. hist. studies of geog. methodology and illustrative works); (2) geog. thought in the 20th century (by topics, incl. genl. surveys, methodology of geog. in genl., theory of regions, landscapes, and boundaries, each subcl. by countries); 400 entries; auth. and subj. indexes.

Pfeifer, Gottfried. *Regional Geography in the United States Since the War: A Review of Trends in Theory and Method.* Translated from the German by John Leighly. New York, Amer. Geog. Soc., 1938. 37 mimeo. pp. [163]
Translation of a paper originally published in *Zeitschrift der Gesellschaft für Erdkunde zu Berlin* [716], 1938, pp. 93–125 (reviewed by John Leighly in *Geog. Rev.* [476], 28 (1938), 679). "Pfeifer has listed a generous share of the literature bearing on his problem in bibliographic notes that can serve as a sufficient guide to recent American geography" (Leighly, *loc. cit.*). Note 1, in particular, lists a selection of works dating from 1906 to 1936 dealing in the large with the progress and methodology of geog. studies in the United States.

Obruchev, V. A., ed. *Uspekhi geologo-geograficheskikh nauk v S.S.S.R. za 25 let: The Progress of the Geological and Geographical Sciences in the U.S.S.R. for 25 Years.* (Academy of Sciences of the U.S.S.R.: Department of Geological and Geographical Sciences.) Moscow and Leningrad, 1943. 200 pp. [164]
In Russian with English title page and table of contents. Deals primarily

with geology. Chapter on geog. by A. A. Grigoriev, pp. 179–197. See also [887].

La Science française. New ed. Paris, Larousse [1933]. 2 vols.   [165]
Broad surveys of the history and status of the several sciences in France, each followed by a bibl., incl. Emmanuel de Martonne, "La Science géographique," in Vol. 1, pp. 373–396. Other chapters of geog. interest deal with geology, ethnology, anthropology, "l'Américanisme et la France," prehistory, oriental studies, and history.

Tulippe, O. *La Géographie dans les universités allemandes.* Liège, J. Wyckmans, 1930. 70 pp. (*Cercle des Géographes Liègeois Fasc. 5 des Travaux, et Travaux du Séminaire de Géographie de l'Université de Liège, Fasc. XXV.*)   [166]
Sections on (1) geog. science in Germany; (2) education for secondary teaching of geog.; (3) geog. institutes (with data on libraries connected therewith), geog. societies, museums. A few bibl. refs. in the footnotes.

HISTORY OF TRAVEL AND EXPLORATION

Bibliographies covering voyages and travels in general: Lewin [337], Vol. 2 (1931), pp. 573–587 (specific voyages); pp. 588–591 (history); pp. 592–652 (collections).   [167]

Cox, E. G. *A Reference Guide to the Literature of Travel, Including Voyages, Geographical Descriptions, Adventures, Shipwrecks, and Expeditions.* Seattle, Univ. of Washington. Vol. 1, *The Old World,* 1935, 410 pp. Vol. 2, *The New World,* 1938, 598 pp. (*Univ. of Washington Publs. in Language and Literature,* Vols. 9 and 10.)   [168]
Covers for period through 1800 books printed in Great Britain, together with "translations from foreign tongues and Continental renderings of English works." By areas and by topics, incl. bibls., collections, circumnavigations, directions for travelers, geog. (prior to 19th century), navigation, maps and atlases, fictitious voyages, etc.; ann.; index of personal names. Review by G. B. Parks in *Geog. Rev.* [476], 29 (1939), 349.

Wright, J. K. "Some Broader Aspects of the History of Exploration," *Geog. Rev.* [476], 25 (1935), 317–320 (review art. on about 15 recent books).   [169]

COLLECTED SOURCES

The Hakluyt Society, London. *Publications,* 1847–.   [170]
"The Hakluyt Society established in 1846, has for its object the printing of rare and valuable Voyages, Travels, Naval Expeditions, and other geographical records." Its publs. comprise some 210 vols. to date (see the Hakluyt Society, *Prospectus* . . . , Cambridge, England, 1938).

Linschoten-Vereeniging, The Hague. *Werken,* 1909–[1939].   [171]
25 parts (*Deel*) in some 46 bound vols. have appeared to date. Texts of Dutch land and sea voyages with bibls., notes, and reproductions of maps. Index to Parts 1–25 (D. Sepp, *Tresoor der zee- en landreizen,* The Hague, Nijhoff, 1939).

The Cortés Society. *Documents and Narratives Concerning the Discovery and Conquest of Latin America.* New York, 1917–1922, 5 nos. in 7 vols.; New Ser., Berkeley, Cal., 1942, 2 nos. in 2 vols. [172]

The Champlain Society. Toronto. *Publications,* 1907–.   [173]
Some 36 vols. to date of texts and translations of early works on the history and exploration of northern North America.

HISTORY OF CARTOGRAPHY

Bibliographical data on the history of cartography: Eckert [99], 1921, 1925.—See also [97].   [174]

Kubitschek, Wilhelm, "Karten," in Pauly [139], 1st Ser., Vol. 10, 1917, pp. 2022–2149.   [175]
Detailed survey of ancient cartography with many bibl. data.

Ireland, H. A. "History of the Development of Geologic Maps," *Bull. Geological Soc. of Amer.,* 54 (1943), 1227–1280.   [176]
Contains many bibl. refs.

Periodical

Imago Mundi: A Periodical Review of Early Cartography
London, 1935–[1939]. Biennial. Ed. by Leo Bagrow and Edward Lyman. Vol. 1, 1935, ed. by Leo Bagrow and Hans Wertheim, published in Berlin, with subtitle *Jahrbuch der alten Kartographie.*
                                                            [177]

Nordenskiöld, A. E. (a) *Facsimile-Atlas to the Early History of Cartography with Reproductions of the Most Important Maps Printed in the XV and XVI Centuries*. Translated from the Swedish original by J. A. Ekelöf and C. R. Markham. Stockholm, 1889. 51 uncolored pl.; 84 uncolored maps in text; 147 pp. text, index, etc. 38 x 52 cm. (b) *Periplus: An Essay on the Early History of Charts and Sailing-Directions*. Translated from the Swedish original by F. A. Bathev. Stockholm, Norstedt, 1897. 60 uncolored pl.; 100 uncolored maps in text; 217 pp. text, index, etc. 38 x 52 cm. [178] The 2 works form a monumental contribution to the history of cartography and geog. (a) deals with printed maps of the period 1475–1592, (b) with both manuscript and printed coast charts from ancient Greek times to the close of the 18th century, with special emphasis on the 14th, 15th, and 16th centuries.

Wieder, F. C., ed. *Monumenta cartographica: Reproductions of Unique and Rare Maps, Plans and Views in the Actual Size of the Originals; Accompanied by Cartographical Monographs*. The Hague, Nijhoff, 1925–1933. 5 vols. 57 x 77 cm. [179]

Humphreys, A. L. *Old Decorative Maps and Charts*. With illustrations from engravings in the MacPherson Collection, and a catalogue of the atlases, etc., in the Collection, by Henry Stevens. London, Halton & Truscott Smith; New York, Minton, Balch Co., 1926. 79 pls. 94 pp. text. 26 x 32 cm. [180] Designed primarily as a guide for the collector of decorative maps. Reviewed in *Geog. Rev.* [476], 18 (1928), 339–340.

Kamal, Prince Youssouf. *Monumenta cartographica Africae et Aegyti*. Cairo, Privately printed, 1926–[1938]. In course of publ. (4 tomes in 13 fascicules have appeared to date). 59 x 76 cm. [181] This enormous work, sponsored by Prince Youssouf Kamal and edited by Professor F. C. Wieder, contains magnificent reproductions of maps dealing with the historical geog. of Egypt and Africa from early Greek times to the Age of Discovery, together with the texts of works bearing on the subject. Ed. limited to 100 copies. Reviewed in *Geog. Rev.* [476], 24 (1934), 175–176.

Miller, Konrad. *Mappaemundi: Die ältesten Weltkarten*. Stuttgart, Roth, 1895–1898. 6 vols. 23 x 29 cm. [182]

Reproductions of medieval world maps with transcriptions and explanatory text.

Miller, Konrad. *Mappae arabicae: Arabische Welt- und Länderkarten des 9.-13. Jahrhunderts in arabischer Urschrift, lateinischer Transkription und Übertragung in neuzeitliche Kartenskizzen, mit einleitenden Texten.* Stuttgart, Published by the editor, 1926-1931. 6 vols. in 3 folios. 25 x 33 cm.   [183]

Hulbert, A. B., ed. *The Crown Collection of Photographs of American Maps.* Cleveland, Ohio, Privately printed by Arthur H. Clark Co., 1904-1928. 5 vols., with index. 43 x 47 cm.   [184]
Photographs of hitherto unpublished maps in the British Museum and other foreign archives especially chosen and prepared to illustrate the early history of the United States and Canada. Edition limited to 25 sets.

Fite, E. D., and Archibald Freeman. *A Book of Old Maps Delineating American History from the Earliest Days Down to the Close of the Revolutionary War.* Cambridge, Harvard Univ. Press, 1926. 75 pp. half-tone reproductions of maps, 293 pp. explanatory text; 20 pp. contents, index, etc. 29 x 42 cm.   [185]
Selection of old maps reproduced to illustrate different phases of American history. Reviewed in *Geog. Rev.* [476], 18 (1928), 339.

Wroth, L. C. "The Early Cartography of the Pacific," *Papers Bibl. Soc. of Amer.,* 38, No. 2 (1944), 85-268.   [186]
Contains a bibl. list of 22 maps reproduced in text and 82 other important maps. Reviewed in *Geog. Rev.* [476], 35 (1945), 505-506.

GEOGRAPHICAL EDUCATION AND METHODOLOGY

(See also [6, No. 28], [160-166].)

Education: Mudge-Winchell [1], bibliographies, encyclopedias, etc., a 162-166, b 26, c 31-33, d 30-31; periodical indexes, a 13, b 10, c 10, d 8.—Bibliographical references to works on and aids to geographical education: Wagner [60], 1938, pp. 24-27.—For periodicals devoted primarily to geographical education see [478, 629, 672, 682, 702, 721, 846, 896]; for bibliographies for teachers of geography, see [22-29].   [187]

Whipple, G. M., ed. *The Teaching of Geography.* Bloomington, Ill.,

Public School Publishing Co., 1933. (*The Thirty Second Yearbook of the National Society for the Study of Education.*) [188]
Coöperative work by many authors. Contains following bibl. chapters: A. P. Brigham and R. E. Dodge, "Nineteenth Century Textbooks of Geography," pp. 3–27; F. K. Branom, "A Bibliography of Geography Books for Teachers and Pupils," pp. 407–413; Elizabeth T. Platt, "Published Materials for the Teachers College Library," pp. 415–427; Norah E. Zink, "Eighty-Two Studies in the Teaching of Geography Classified by Content and Technique, with Selected Summaries," pp. 431–473.

Clark, Rose B. "Geography in the Schools of Europe," Bowman [261], 1934, pp. 229–366. [189]
Bibl. on pp. 345–355.

PHYSICAL AND MATHEMATICAL GEOGRAPHY

General Aids to Study of Natural Science and Mathematics

Science in general: Mudge-Winchell [1], bibliographies, a 169–170, c 35, d 32; dictionaries, a 168–169, c 34, d 32; engineering, bibliographies, b 31–32, c 40–41; engineering, periodical indexes, a 16–17; particular sciences, a 170–192, b 27–29, c 35–38, d 33–36; periodical indexes, a 15–16; tables, a 170, b 27, c 34, d 32–33. [190]
For a record of the most recent progress of investigation in the exact and natural sciences and for reviews and abstracts of the more striking and original books and other publications the well-known weeklies, *Science* (American) and *Nature* (British), are indispensable. These contain news of exploring expeditions and important geographical researches that are in progress or planned for the immediate future. Continental counterparts of these two periodicals are *Die Naturwissenschaften* and *Revue générale des sciences pures et appliquées.* [191]

Catalogue of Scientific Papers, 1800–1900. (Royal Society of London.) London, Clay, 1867–1902; Cambridge Univ. Press, 1914–1925. 19 vols. in 4 ser. *Subject Index,* Cambridge Univ. Press, 1908–1914. 3 vols. in 4. [192]
A vast alph. auth. index to the scientific publs. of the 19th century. Of the subj. index only the vols. for pure mathematics, mechanics, and physics have been issued.

International Catalogue of Scientific Literature. Published for the International Council by the Royal Society of London. London, Harrison, 1901–1919. Annual. [193]
An outgrowth of the *Catalogue of Scientific Papers* [192]. Each annual issue consists of 17 lettered vols. on the various sciences, incl. (F) Meteorology, (H) Geology, (J) Geography (Mathematical and Physical), (L) General Biology, (P) Physical Anthropology. (J), covering 1901–1916, lists both books and period. arts. alph. by auths., with system. subj. catalogue and alph. subj. indexes.

Mathematical Geography, Geodesy, Photogrammetry

(See also [92–131], cartography and atlases.)
Bibliographical data on mathematical geography: Wagner [60], 1938, pp. 66–68, and footnotes on pp. 68–267.—For a periodical devoted primarily to mathematical geography, surveying, cartography, and photogrammetry see [844].—Navigation, tables and other reference works: Mudge-Winchell [1], a 173–174. [194]

Perrier, Georges, and others. *Bibliographie géodésique internationale.* (Association de Géodésie de l'Union Géodésique et Géophysique Internationale.) Paris, Secrétariat de l'Association. Vol. 1, *Années 1928 . . . 1930, 1935,* 219 pp. Vol. 2, *Années 1931 . . . 1934, 1938,* 420 pp. [195]
Main bibl. cl. by topics; 1,433 and 5,554 entries; ann.; auth. indexes; list of periodicals by countries.

American Society of Photogrammetry. *Bibliography of Photogrammetry,* constituting *Photogrammetric Engineering,* Vol. 2, No. 4, Oct.–Dec., 1936. 117 pp. [196]
Covers books, period. arts., and govt. publs.; alph. by auths., topical index.

Gore, J. H. "A Bibliography of Geodesy," *U.S. Coast and Geodetic Survey: Report for 1887,* Appendix No. 16, Washington, 1889, pp. 313–512. A second edition was published as Appendix No. 8 of the *Report* for 1902, Washington, 1903, pp. 427–787. [197]
Alph. by auths. and topics.

114   *Topical Aids*

Physical Geography and Related Sciences

(See also [6, Nos. 6–7].)

Atlas

Berghaus' physikalischer Atlas. 3d ed. Ed. by Hermann Berghaus. Gotha, Justus Perthes, 1887–1892. 75 pl.; 82 pp. text, contents, index. 27 x 39 cm.   [198]
Geology, hydrography and oceanography, meteorology, terrestrial magnetism, flora, fauna, ethnography.

GEOLOGY AND GEOCHRONOLOGY

(See also [6, No. 8]; [176], maps; [321], military geology.)

According to the survey recently made by the American Library Association (see above, p. 23, note 20), the following American libraries contain important collections of geological publications:
Universities: Alabama, California, Chicago (paleontology), Columbia (physiography), Cornell, Harvard, Johns Hopkins, Joint University Libraries (Nashville, Tenn.), Massachusetts Institute of Technology, Michigan, Minnesota, North Carolina, Ohio State, Oklahoma, Princeton (vertebrate paleontology), Stanford, Washington at Seattle, Yale. *Public Library:* New York; *Others:* American Museum of Natural History (vertebrate paleontology), Engineering Societies Library, Library of Congress, Philadelphia Academy of Natural Sciences, Smithsonian Institution, United States Geological Survey.   [199]

BIBLIOGRAPHIES OF BIBLIOGRAPHIES

Mathews, E. B. *Catalogue of Published Bibliographies in Geology, 1896–1920.* Washington, 1923. 228 pp. (*Bull. of the National Research Council,* No. 36 (Vol. 6, Pt. 5).)   [200]
Continuation of [201]. 3 main parts: (1) genl. bibls.; indexes of series, etc.; (2) subj. bibls. (alph. by topics and areas); (3) personal bibls. and necrologies. 3,312 entries; auth. index.

Margerie, Emmanuel de. *Catalogue des bibliographies géologiques, rédigé avec le concours des membres de la Commission Biblio-*

graphique du Congrès. (Congrès Géologique International.) Paris, Gauthier-Villars, 1896. 753 pp. [201]
A fundamental work. Topical and areal cl.; 3,918 entries; ann.; indexes by auths., topics, and geog. names. For continuation see [200].

CURRENT BIBLIOGRAPHIES

Annotated Bibliography of Economic Geology. Urbana, Ill., Economic Geology Publishing Co. 1928–. Semiannual. Prepared under the auspices of the Society of Economic Geologists. Genl. index to first 10 vols., 1928–1938. [202]

Bibliography and Index of Geology Exclusive of North America. [New York], Geological Soc. of Amer. 1933–. Annual (1 vol. for 1941–1942). [203]
Compiled by J. M. Nickles and others. Complement to United States Geological Survey's bibls. of geology of North America [427]. Supersedes [204]. In each vol. main bibl. alph. by auths. with detailed topical and place index; also list of periodicals.

List of Geological Literature Added to the Geological Society's Library During the Year . . . London, Geological Society. 1895–1936. Annual. [204]
Comprehensive lists, alph. by auths., with subj. index. Superseded by [203].

OTHER REFERENCE WORKS

Mudge-Winchell [1], a 181–184, b 29, c 36–37, d 34. [205]

Internationaler Geologen- und Mineralogen-Kalender. Stuttgart, Ferdinand Enke. ?–[1937] Irregularly. Ed. by Deutsche Geologische Gesellschaft, Berlin. Formerly *Geologen-Kalender.* [206]
Lists geologists and also geological institutions, university departments, etc., with data on their publs.

Fleming, R. C. *Source Book: A Directory of Public Agencies in the United States Engaged in the Publication of Literature on Mining and Geology.* New York, S.W. Mudd Memorial Fund, Amer. Inst. of Mining and Metallurgical Engineers, 1933. 128 pp. [207]

GEOCHRONOLOGY

A series of annotated bibliographies by J. P. Marble of articles relating to the measurement of geologic time have been published since

1936 in the mimeographed reports of the Committee on the Measurement of Geologic Time of the National Research Council, Washington, D.C.   [208]

Schulman, Edmund. "A Bibliography of Tree-Ring Analysis," *Tree-Ring Bull.,* 6 (1940), 27–39.   [209]
412 entries.

Arctowski, Henryk. *A Bibliography of Scientific Papers on Climatic Variations.* (International Geographical Union: Commission of Climatic Variations.) Lwów, 1938. 254 mimeo. pp.   [210]
Alph. by topics; 4,153 entries; auth. index.

GEOMORPHOLOGY

(See also [6, Nos. 6, 7, 9]; [321], military geomorphology.)

Bibliographical data on submarine canyons: note in *Geog. Rev.* [476], 27 (1937), 681–683.   [211]

Bibliographie des cartes de surfaces d'aplanissement. (Union Géographique Internationale: Commission pour la Cartographie des Surfaces d'Aplanissement.) Paris, [1938]. 79 pp.   [212]
Alph. by auths.; 272 entries; index to geog. names.

Maull, Otto. *Geomorphologie.* Leipzig and Vienna, Deuticke, 1938.
(*Enzyklopädie der Erdkunde.*)   [213]
Bibl. on pp. 423–480: topical cl., 756 entries.

Spreitzer, Hans. "Die Fortschritte der Geomorphologie (Exogene Kräfte und ihre Wirkungen) (1925–36)," *Geog. Jhrb.* [6], 52 (1937), 415–476; 53, (1938), 1, 3–254.   [214]
2 main parts: (1) genl., (2) geomorphic forces and resulting landforms; topical cl., some topics subcl. by areas; 3,012 entries; ann.; auth. index. Page refs. are given under the appropriate hdgs. in Part III, below, to the section on "comprehensive geomorphic studies of particular areas"; many of the other sections also contain refs. for the same areas.

Jung, H., and G. Selzer. "Endogene Vorgänge und Formbildung (1927–35), mit einem Anhang: Physikalische Untergrundforschung (angewandte Geophysik)," *Geog. Jhrb.* [6], 52 (1937), 249–414.   [215]
By topics, incl. form, rotation, age, temperature, etc., of earth; tectonics, vulcanism, tides, gravity, earthquakes, geophysical methods; 2,095 entries; ann.; auth. index.

Williams, G. R., and others. *Selected Bibliography on Erosion and Silt Movement*. Washington, 1937. 91 pp. (*U.S. Geological Survey Water-Supply Paper* 797.) [216]

Haferkorn, H. E. *Sand Movements and Beaches, Including References on Bars, Bays, Coast Changes, Currents, Erosion, Estuaries, Shorelines, Tides, Waves and Wave Action. . . . A Bibliography.* Prepared under the direction of the Chief of Engineers, U.S. Army. Fort Humphreys, Va., The Engineer School. 1929. 124 pp. [217]
Topical cl.; 1,101 entries; auth. index.

Periodicals

Revue de géographie physique et de géologie dynamique: Bulletin du Laboratoire de Géographie Physique de la Faculté des Sciences de l'Université de Paris
Paris, Les Presses Universitaires de France, 1928–[1939]. Quarterly. Ed. under the direction of Léon Lutaud and Jacques Bourcart. [218]
Substantial original material; of less value as bibl. tool.

Journal of Geomorphology
New York, Columbia Univ. Press, 1938–[1942]. Quarterly. Ed. by the late Professor Douglas Johnson. [219]
Substantial arts. and critical reviews. (See *Geog. Rev.* [476], 28 (1938), 503).

SEISMOLOGY

Hodgson, E. A. "Bibliography of Seismology," *Publs. of the Dominion Observatory,* Ottawa. 1929–. Irregularly. [220]
A few anns.

Hodgson, E. A. "Bibliography of Seismology," *Bull. Seismological Soc. of Amer.,* 17 (1927), 149–182, 218–248; 18 (1928), 16–63, 110–125, 214–235, 267–283; 19 (1929), 206–227. [221]
Alph. by auths., 1,200 entries; ann. Superseded by [220].

Davison, Charles. *Studies on the Periodicity of Earthquakes*. London, Murby, 1938. [222]
List of 131 catalogues of earthquakes on pp. 102–107.

Montessus de Ballore, Fernand de. *Bibliografía general de temblores y terremotos*. (Sociedad Chilena de Historia y Geografía.) Santiago de Chile, Imprenta Universitaria, 1915–1919. 7 pts. 1,515 pp. (Also published in *Revista Chilena de Historia y Geografía* [575], 1915–1919.) [223]

TERRESTRIAL MAGNETISM

Fleming, J. A., ed. *Terrestrial Magnetism and Electricity.* New York and London, McGraw-Hill, 1939. (*Physics of the Earth,* Vol. VIII.) [224]
"Bibliographical Notes and Selected References" by H. D. Harradon on pp. 679–778; topical cl.; 1,523 entries.

Bartels, J. "Bericht über die Fortschritte unserer Kenntnisse vom Magnetismus der Erde (IX, 1925–29)," *Geog. Jhrb.* [6], 44, (1929), 3–36. [225]
2 parts: (1) genl., subcl. topically; (2) observatories, surveys, field studies, subcl. areally; 417 entries; ann.; auth. index.

SNOW AND ICE

A list of current publications on snow and ice has appeared since 1935 in Part 2 of the *Trans. of the American Geophysical Union* published annually by the National Research Council, Washington.—See also [1172]. [226]

Bibliography on Ice of the Northern Hemisphere. (United States Navy Department: Hydrographic Office.) Washington, 1945. 191 pp. (*H.O. Pub.,* No. 240) [227]
The bibl. for [227a]. Compiled by Mary C. Grier. Cl. by bodies of water (e.g. Arctic Ocean, Greenland Sea) with sections on rivers and lakes of North America and of Eurasia; 1,700 entries; indexes by subjects, names (auths., ships, expeditions, etc.), geog. names.

Maps and Atlas

Ice Atlas of the Northern Hemisphere. (United States Navy Department: Hydrographic Office.) Washington, 1946. 106 pp. 62 x 62 cm. (*H. O. Pub.* No. 550.) [227a]
Compiled under the direction of J. C. Weaver. Covers sea and river ice for Northern Hemisphere as a whole and sea ice for Grand Banks region, Baltic, Black, and White seas, and Okhotsk Sea regions. Bibl. on pp. 93–106: 1,700 entries; no index (see, however, [227]).

Antevs, Ernst. "Maps of the Pleistocene Glaciations," *Bull. Geological Soc. of Amer.*, 40 (1929), 631–720.   [228]
Bibl. on pp. 696–720.

SOILS

(See also [299], atlas.)

Review of four publications on tropical soils by R. L. Pendleton: *Geog. Rev.* [476], 29 (1939), 493–496.   [229]

Giesecke, F. "Die bodenkundliche Forschung, 1927/28–1937: Ein Überblick," *Geog. Jhrb.* [6], 54, Pt. 1 (1939), 181–302.   [230]
Areal cl. with introductory section cl. by topics; 2,111 entries; ann.; auth. index.

WATERS OF THE LAND

(See also [6, No. 10], [235].)

Friedrich, Wilhelm. "Gewässerkunde (1930–37)," *Geog. Jhrb.* [6], 54, Pt. 1 (1939), 85–180.   [231]
Topical cl. with areal subcl. for certain topics; 1,125 entries; ann.; auth. index.

Chumley, James. "Bibliography of Limnological Literature," in Sir John Murray and Laurence Pullar, *Bathymetrical Survey of the Scottish Fresh-Water Lochs,* Edinburgh, 1910, Vol. 1, pp. 659–753.   [232]
Alph. by auths.; no anns. or index.

OCEANOGRAPHY

(See also [6, No. 11], [259], animal geography.)

Lewis, C. L. *Books of the Sea: An Introduction to Nautical Literature.* Annapolis, U.S. Naval Inst., 1943. 318 pp.   [233]
Each chapter is a running discussion of some phase of nautical literature (incl. fiction, poetry, plays, books of travel, naval histories, works on oceanography, etc.) followed by "reading lists." Refs. are preponderantly to works in English.

A bibliography on tides and kindred matters, covering the period 1910–1937 with certain publications before 1910, has been published in five installments, (1) and (2) comprising *Union Géodésique et Géophysique Internationale: Section d'Océanographie, Bulletin*, Nos. 12 and 17; (3), (4), and (5) comprising *Publications scientifique de l'Association d'Océanographie Physique*, Nos. 2, 3, and 6. [234]

Bibliographia oceanographica: Johannes Magrini, Fundator. Rome (formerly Venice), Consiglio Nazionale Ricerche. 1930–[1939]. Annual vols. issued in parts. [235]
Extensive ann. bibl. covering oceanography, lakes, rivers, fisheries, and navigation. Preceded by Giovanni Magrini, *Essai d'une bibliographie générale des sciences de la mer* . . . *année 1928*, Venice, Ferrari, 1929.

Schulz, Bruno. "Bericht über die Fortschritte der Ozeanographie (1933–37)," *Geog. Jhrb.* [6], 54, Pt. 1 (1939), 3–84. [236]
Deals primarily with physical oceanography. 2 parts, (1) topical cl., (2) by areas subcl. by topics; 1,154 entries; ann., auth. index.

Index to Publications by the International Council for the Exploration of the Sea, 1899–1938. (Conseil Permanent International pour l'Exploration de la Mer.) Copenhagen, Høst, 1939. 145 looseleaf pp. [237]
In 8 parts: (1) alph by auths., etc.; (2) administration and economics; (3) hydrography; (4) meteorology; (5) the sea bottom; (6) plankton; (7) biology; (8) fisheries.

Marmer, H. A. "Recent Major Oceanographic Expeditions: A Review of the Work of the Meteor, Carnegie, Dana, and Snellius Expeditions," *Geog. Rev.* [476], 23 (1933), 299–305. [238]

Tanfil'ev, G. Ī. *Moria: Kaspīĭskoe, Chernoe, Baltīĭskoe, Ledovitoe, Sibīrskoe, ī Vostochnyĭ Okean: Īstoriia Issledovaniia: Morfometriia, Gīdrologiia, Bīologiia*. [Seas: Caspian, Black, Baltic, Arctic, Siberian, and Pacific Ocean: History of Exploration: Morphometry, Hydrology, Biology]. Moscow and Leningrad, 1931. 248 pp. [238a]
Comprehensive bibls. at ends of chapters.

Bencker, H. "The Bathymetric Soundings of the Oceans," *Hydrographic Review*, 7 (1930), 64–97. [239]
Contains a chronological list (without bibl. refs., however) of ocean explorations 1800–1930 (pp. 72–86).

METEOROLOGY AND CLIMATOLOGY

(See also [6, No. 12].)

Meteorology, bibliographies, and other reference works: Mudge-Winchell [1], a 184–185, c 37, d 34.—Note on bibliographies of general meteorology in U.S. Dept. of Commerce, Weather Bureau, *Library Circular,* July, 1943, pp. 1–5. [240]

Bibliography of Meteorological Literature Prepared by the Royal Meteorological Society with the Collaboration of the Meteorological Office. London, Royal Meteorological Soc., 1920–. Semiannual. [241]
Succeeds bibls. in *Quarterly Jour. of the Royal Meteor. Soc.,* 1917–1920.

Muller, Hans. "On International Bibliographies of Meteorology," *Bull. Amer. Meteorological Soc.* 23 (1942), 407–410. [242]
A survey of "serially issued bibliographies of international scope."

Köppen, W., and R. Geiger, eds. *Handbuch der Klimatologie.* Berlin, Bornträger. 1930–[1939]. 5 vols. [in 14 parts]. [243]
Bibls. and bibl. footnotes are included in most of the parts of this fundamental series, which covers to date general climatology (Vol. 1), the Americas (Vol. 2), Europe and U.S.S.R. in Asia (Vol. 3), Farther India, Insulindia, Southern Australia, New Zealand, South Sea Islands, and Antarctica (Vol. 4), and Rhodesia, Nyasaland, and Mozambique Colony (Vol. 5). The parts on the West Indies, North America, Australia, New Zealand, and Rhodesia, etc., are in English, the others in German. See reviews in *Geog. Rev.* [476], 21 (1931), 527; 22 (1932), 161, 511; 27 (1937), 702; 28 (1938), 695.

United States Works Progress Administration. *Bibliography of Aeronautics:* Part 2, *Meteorology.* Compiled from the *Index of Aeronautics* of the Institute of the Aeronautical Sciences. New York, 1937. 641 mimeo. pp. [244]
Topical cl., auth. index.

Ward, R. DeC. "The Literature of Climatology," *Annals Assn. of Amer. Geogs.* [479], 21 (1931), 34–51. [245]
Not strictly a bibl. but a "stock taking" of "the main lines along which climatological investigation and writing has progressed . . . in order that those who are primarily concerned with climatological literature may have this matter brought somewhat more clearly to their attention."

Knoch, K. "Bericht über die Arbeiten aus dem Gebiete der geographischen Meteorologie, 1926–28," *Geog. Jhrb.* [6], 44 (1929), 37–144.   [246]
3 main parts: (1) genl., (2) genl. climatology, (3) descriptions of climates. (1) and (2) cl. by topics, (3) by regions; 1,296 entries; ann., auth. index.

Atlases

Atlas of Climatic Charts of the Oceans. Prepared under the supervision of W. F. McDonald. Washington, 1938. 130 charts (2 to a pl.); 6 pp. text, etc. 48 x 46 cm. (U.S. Weather Bur. [Publ.] No. 1247.)   [247]
Charts (Mercator's projection), based on observations taken by ships at sea during more than fifty years, covering winds (48 charts), fog, mist, haze, exceptional visibility (16), clouds (28), precipitation (16), mean depression of the wet-bulb thermometer (4), temperature of air and sea surface (16). Reviewed by John Leighly in *Geog. Rev.* [476], 30 (1940), 175–176.

Bartholomew, J. G., and A. J. Herbertson. *Atlas of Meteorology.* Ed. by Alex. Buchan. (Royal Geog. Soc.) London, Constable, 1899. 34 pl. with more than 400 maps; 57 pp. text, appendices, index, etc. 31 x 47 cm. (*Bartholomew's Physical Atlas,* Vol. 3.)   [248]
2 parts: (1) climate maps (temperature, pressure, winds, clouds, precipitation), (2) weather maps (storms, anomalous and typical weather).

GEOGRAPHY OF PLANTS AND ANIMALS (BIOGEOGRAPHY)

Biology: bibliographies and other reference works: Mudge-Winchell [1], a 186–187.   [249]

Plant Geography (Phytogeography)

(See also [6, No. 13].)

Botany: bibliographies and other reference works: Mudge-Winchell [1], a 187–189, c 37, d 35.   [250]

Cain, S. A. *Foundations of Plant Geography.* New York, Harper, 1944.   [251]
Bibl. of "literature cited" on pp. 491–528: alph. by auths.; 720 entries.

Blake, S. F., and Alice C. Atwood. *Geographical Guide to Floras of the World: An Annotated List with Special Reference to Useful Plants and Common Plant Names.* Part I, *Africa, Australia, North America, South America, and Islands of the Atlantic, Pacific, and Indian Oceans.* Washington, 1942. 336 pp. (U.S. Dept. of Agric., *Miscellaneous Publ.,* No. 401.)   [252]
"Annoted catalog of all the more useful floras and floristic works, including those in periodical literature, that list or describe the complete vascular flora . . . of any region or locality . . . [aims to] include as well all publications dealing on the same scale with useful and medicinal plants, vernacular names, and botanical bibliography." Areal cl.; auth. and place indexes.

Raup, H. M. "Trends in the Development of Geographic Botany," *Annals Assn. of Amer. Geogs.* [479], 32 (1942), 319–354.   [253]
"Literature cited," pp. 350–354 (alph. by auths.).

Weaver, J. E., and F. E. Clements. *Plant Ecology.* 2d ed. New York and London, McGraw-Hill, 1938.   [254]
Bibl. on pp. 539–582: alph. by auths., 1,035 entries.

Diels, L. "Pflanzengeographie (1927–35)," *Geog. Jhrb.* [6], 51, (1936), 200–229.   [255]
Areal cl., with introductory topical section; 467 entries; ann.; auth. index.

Animal Geography (Zoogeography)

(See also [6, No. 15].)

Zoology: bibliographies and other reference works: Mudge-Winchell [1], a 190–192, d 35–36.   [256]

Murphy, R. C. "Animal Geography: A Review," *Geog. Rev.* [476], 28 (1938), 140–144.   [257]
Deals with about 12 works.

Rensch, Bernhard. "Tiergeographie (1931–1937)," *Geog. Jhrb.* [6], 53, Pt. 2 (1938), 369–436.   [258]
4 parts: (1) genl.; (2) ecological; (3) regional; (4) marine; (1) and (2) cl. by topics, (3) and (4) by areas; 668 entries; ann.; auth. index.

Ekman, Sven. *Tiergeographie des Meeres.* Leipzig, Akademische Verlagsgesellschaft, 1935.   [259]
Bibl. on pp. 480–505 (alph. by auths.).

HUMAN GEOGRAPHY AND RELATED FIELDS

(See also [6, Nos. 16–19], [1161], population and settlement.)

General and Miscellaneous Works

Social sciences: Mudge-Winchell [1]: bibliographies, a 113–114, b 20; encyclopedias, a 113, b 20; periodical indexes, a 16. [260]

Bowman, Isaiah. *Geography in Relation to the Social Sciences.* (American Historical Association.) New York, Scribner's, 1934. (*Report of the Commission on the Social Studies,* Part 5.) [261]
Suggestive bibl. refs. in footnotes. See also [189].

Ellwood, C. A., and others. *Recent Developments in the Social Sciences.* Philadelphia and London, Lippincott, 1927. (*Lippincott's Sociological Series.*) [262]
Collaborative work with chapters on sociology, anthropology, psychology, cultural geog. (by C. O. Sauer, pp. 154–212), economics, political science, and history. Professor Sauer's chapter is well documented with bibl. refs.

Brunhes, Jean. *Human Geography: An Attempt at a Positive Classification, Principles and Examples.* Transl. by T. C. Le Compte, ed. by Isaiah Bowman and R. E. Dodge. Chicago and New York, Rand McNally, 1920. 664 pp. [263]
The index shows that more than 500 authors are represented in the footnote references in this study. A selected list of the more important works published since 1925 is to be found in Brunhes' *La Géographie humaine,* 4th ed., Vol. 2, 1934, pp. 977–987. There are also many footnote refs. in his survey of "Human Geography" published in *The History and Prospects of the Social Sciences,* ed. by H. E. Barnes (New York, Knopf, 1925), pp. 55–105.

Ethnology: bibliographies and other reference works: Mudge-Winchell [1], a 185–186, d 34–35. [264]
The Cross-Cultural Survey, established by the Institute of Human Relations of Yale University in 1937 under the supervision of Professor G. P. Murdock, "has assembled and organized the available information on about 150 peoples—mainly but not exclusively primitive—in all parts of the world. All the relevant data were abstracted in full from the sources, translated from foreign languages, classified

according to subject, and transferred to files. This complete file is maintained at Yale University." [265]

Grau, Rudolf. "Die völkerkundliche Forschung (1909–31)," *Geog. Jhrb.* [6], 47 (1932), 271–348; 48 (1933), 3–50. [266]
Areal cl., with sections on bibls., periodicals (299 titles), and genl. works; 2,590 entries; ann.; auth. index. See also [6, No. 16].

———

Odum, H. W., and H. E. Moore. *American Regionalism: A Cultural-Historical Approach to National Integration.* New York, Holt, 1938. [267]
The footnotes and the bibl. on pp. 643–675 provide refs. to works on regionalism in genl. and on its manifestations in the United States.

———

Medicine: Mudge-Winchell [1]: bibliographies and other reference works, a 196–198, b 30, c 39, d 37–38; periodical indexes, a 14–15, b 10, c 10, d 9.—The number of *Ciba Symposia* (published by Ciba Pharmaceutical Products, Inc., Summit, N.J.) for January, 1945 (Vol. 6, No. 10), pp. 1986–2020, is devoted to medical geography. It comprises three papers by Arne Barkhuus: "Medical Surveys from Hippocrates to the World Travellers," "Medical Geographies" (a review of the outstanding books published 1746–1944), and "Geomedicine and Geopolitics." [268]

Simmons, J. S., and others. *Global Epidemiology: A Geography of Disease and Sanitation.* Vol. 1, Pt. 1, "India and the Far East"; Pt. 2, "The Pacific Area." Philadelphia, London, Montreal, J. B. Lippincott, 1944. 530 pp. [269]
To be completed in 4 vols. Each chapter deals with a specific region and is divided into sections on geog. and climate, public health, medical facilities, and diseases, followed by a bibl.

———

Kurath, Hans. *Handbook of the Linguistic Geography of New England.* Providence, Brown University, 1939. [270]
Accompanies *Linguistic Atlas of New England.* Bibl. of linguistic geog. on pp. 54–61 covering: (1) the chief bibls., (2) the more important journals and serials, (3) linguistic atlases published or in process of publication, (4) a selection of books and arts.; 204 entries.

Schrijnen, Jos. *Essai de bibliographie de géographie linguistique générale.* Nimègue, 1933. 96 pp. (*Comité International Permanent*

de Linguistes: Publications de la Commission d'Enquête Linguistique, 2.) [271]
Contains a short bibl. (pp. 11–17) and data on linguistic atlases in existence and in preparation.

———

Revue pour l'étude des calamités: Bulletin de l'Union Internationale de Secours. Geneva, 1938–[1945]. Irregularly. Succeeds *Matériaux pour l'étude des calamités,* publ. 1924–1933 by the Société de Géographie de Genève in collaboration with the Comité International de la Croix Rouge and the Ligue des Sociétés de la Croix Rouge; 1934–1945 by the Union Internationale de Secours. [272]
Deals with calamities of all kinds affecting large numbers of people and with means of preventing and ameliorating them. Contains bibls. and book reviews.

———

Beach, H. P., and C. H. Fahs, eds. *World Missionary Atlas.* Containing a directory of missionary societies, classified summaries of statistics, maps showing the location of mission stations throughout the world, a descriptive account of the principal mission lands and comprehensive indices, maps by John Bartholomew. New York, Institute of Social and Religious Research, 1925. 29 pl. 348 pp. text. 26 x 37 cm. [273]

———

Bibliographical data on the geography of rural settlements: note in *Geog. Rev.* [476], 24 (1934), 502–504. [274]

Bibliographical data on geographical aspects of tourism: note in *Geog. Rev.* [476], 25 (1935), 507–509. [275]

Urban Geography

Mumford, Lewis. *The Culture of Cities.* New York, Harcourt Brace, 1938. [276]
Bibl. on pp. 497–552: alph. by auths.; ann.

(a) Kimball, Theodora. *Manual of Information on City Planning and Zoning, Including References on Regional, Rural, and National Planning.* Cambridge, Harvard Univ. Press, 1923. 198 pp. (b) Hubbard, T. K., and Katherine McNamara. *Planning Information Up-to-Date: A Supplement, 1923–1928, to Kimball's Manual of In-*

formation on City Planning. 1928. 111 pp. (c) McNamara, Katherine. *Bibliography of Planning, 1928–1935: A Supplement to Manual of Planning Information . . . , 1928, by T. K. Hubbard and Katherine McNamara.* 1936. 240 pp. (*Harvard City Planning Studies,* Vol. 10. 1936.) [277]
Cover town, city, regional, state, and national planning; although the emphasis is mainly on the United States, each vol. contains a section on planning in other countries. Topical cl. (system.); auth. and subj. indexes. (a) and (b) were also issued in a single bound volume in 1928. (c) was reviewed by C. C. Colby in *Geog. Rev.* [476], 27 (1937), 524–526.

Aurousseau, M. "Recent Contributions to Urban Geography," *Geog. Rev.* [476], 14 (1924), 444–455. [278]
Running discussion of the development of city geog. with numerous refs. in the footnotes.

Economic Geography

(See also [6, No. 18], [28].)

Economics: Mudge-Winchell [1]: encyclopedias, a 114–115; period. index, a 17.—For periodicals devoted primarily to economic geography see [480, 671, 740]. [279]

Bibliographical data on "Conservation, Old and New": note in *Geog. Rev.* [476], 32 (1942), 498–501. [280]

Broek, J. O. M. "Discourse on Economic Geography," *Geog. Rev.* [476], 31 (1941), 663–674. [281]
Review of 12 recent books in form of a dialogue.

United States Tariff Commission. *Raw Materials Bibliography: General References to Selected Raw Materials and Basic Economic Resources.* Washington, 1939. 85 mimeo. pp. [282]
Alph. by auths.; 728 entries; ann.; topical index.

Bengtson, N. A., and William Van Royen. *Fundamentals of Economic Geography.* New York, Prentice Hall, 1935. 830 pp. [283]
Bibl. on pp. 765–783.

Lütgens, Rudolf. "Wirtschaftsgeographie, einschliesslich Verkehrsgeographie (1908–34)," *Geog. Jhrb.* [6], 50 (1935), 135–318; 51 (1936), 3–199. [284]

3 parts: (1) auxiliary works (periodicals, genl. ref. books, atlases and world maps, etc.) and methodology; (2) genl. works; (3) regional; (1) and (2) cl. by topics, (3) by areas; 4,528 entries; ann.; auth. index.

Atlases

Bartholomew, John. *The Oxford Economic Atlas.* 8th ed. London, Oxford Univ. Press, 1937. 64 pp. colored maps. 12 pp. text. 23 x 29 cm. [285]

Baratta, Mario, and Luigi Visintin. *Atlante della produzione e dei commerci.* 3d ed. Novara, 1933. 88 pp. 27 x 34 cm. [286]

Philip, George, and T. S. Sheldrake, eds. *The Chambers of Commerce Atlas: A Systematic Survey of the World's Trade, Economic Resources and Communications.* London, Geo. Philip, 1925. (Association of British Chambers of Commerce.) *Ca.* 225 pl. and diagrams; 49 pp. text, contents, commercial compendium, index, etc. 27 x 40 cm. [287]

AGRICULTURAL GEOGRAPHY

Agriculture: Mudge-Winchell [1]: bibliographies, a 194–195, c 39, d 37; encyclopedias, a 195; libraries, c 81; periodical indexes, a 11–12, b 9, c 9–10, d 8; statistics, a 196, b 30, c 38. [288]

A Survey of Current Bibliographies on Agriculture and Allied Subjects. Rome, International Inst. of Agric., 1937. 84 pp. Title also in French. [289]
Cl. by countries; ann.; topical and title indexes.

Ogden, E. Lucy, and Emma B. Hawks. *List of Manuscript Bibliographies and Indexes in the U.S. Department of Agriculture, Including Serial Mimeographed Lists of Current Literature.* Washington, 1926. 38 mimeo. pp. (U.S. Dept. of Agric., *Bibliographical Contributions No.* 11.) [290]

———

Klages, K. H. W. *Ecological Crop Geography.* New York, Macmillan, 1942. 633 pp. [291]
Bibl. refs. at ends of chapters.

Olcott, Margaret T. *The World Food Supply: A Partial List of References, 1925–1939.* Washington, 1939. 164 mimeo. pp. (U.S. Dept.

of Agric.: Bureau of Agric. Economics, *Agric. Economics Bibl.,* No. 82.) [292]
By continents, subcl. by countries (a few further subcl. by topics), with section on "world" subcl. by topics; genl. index.

Goodsell, O. E. *Land Classification: A Selected Bibliography.* Washington, 1940. 102 mimeo pp. (U.S. Dept. of Agric.: Bureau of Agric. Economics, *Agric. Economics Bibl.,* No. 83.) [293]
"The references selected are concerned in general with classifications which group land areas according to problems for the purpose of recommending certain measures of land-use adjustment. They are not concerned with classifications of land according to types of farming, types of vegetation, types of soil, etc." Areal cl., with genl. section subcl. by topics; auth. index.

Bercaw, Louise O., and Annie M. Hannay. *Bibliography on Land Utilization, 1918-1936.* Washington, 1938. 1,508 pp. (U.S. Dept. of Agric., *Miscellaneous Publ.,* No. 284.) [294]
Companion vol. to [295]. Areal cl., subcl. by topics and areas; fully ann.; genl. index.

Bercaw, Louise O., and others. *Bibliography on Land Settlement: With Particular Reference to Small Holdings and Subsistence Homesteads.* Washington, 1934. 492 pp. (U.S. Dept. of Agric., *Miscellaneous Publ.,* No. 172.) [295]
3 parts: (1) genl., (2) U.S. (subcl. by topics and by states); (3) foreign countries (subcl. by countries); fully ann.; genl. index.

Bibliography of Tropical Agriculture. Rome, International Institute of Agriculture, 1932-1939. Annual. Title also in French. [296]
By topics, incl. specific crops and plants, with a section "Articles on Regions"; auth. index. Abstracts are given for most of the titles.

Pan American Union. *Selected List of Publications on Tropical Agriculture.* Prepared for the Inter-American Conference on Agriculture, Forestry and Animal Industry, September 8-20, 1930. Washington, 1930. 90 mimeo. pp. [297]
By topics, incl. genl. agric., specific plants, crops, etc.

Revue internationale de botanique appliqué et d'agriculture tropicale. Paris, 1920-. Monthly. Published under the direction of August Chevalier. [298]

Atlases

Krische, Paul. (a) *Bodenkarten und andere kartographische Darstellungen der Factoren der landwirtschaftlichen Produktion verschiedener Länder: Ein Beitrag zur neuzeitlichen Wirtschaftsgeographie.* Berlin, Parey, 1928. 111 pp. 77 maps. (b) *Landwirtschaftliche Karten als Unterlagen wirtschaftlicher, wirtschaftsgeographischer und kulturgeschichtlicher Untersuchungen.* Berling, Deutsche Verlagsgesellschaft, 1933. 112 pp. 209 maps. (c) *Mensch und Scholle: Kartenwerk zur Geschichte und Geographie des Kulturbodens.* Berlin, Deutsche Verlagsgesellschaft, 1936. 151 pp. 289 maps. [299] With a few exceptions the maps are either adaptations or direct reproductions of maps that have appeared in many different publs. In (a) the emphasis is mainly on soil maps; in (b) and (c), while there are also soil maps, a much wider range of economic-geog. subjs. is covered. Despite their heterogeneous character these volumes may serve as a useful reference atlas in agricultural geog. Reviewed ((a) and (b) by C. F. Marbut) in *Geog. Rev.* [476], (a), 19 (1929), 345–346; (b), 24 (1934), 174–175; (c), 28 (1938), 174–175.

Finch, V. C., and O. E. Baker. *Geography of the World's Agriculture.* (United States Department of Agriculture.) Washington, 1917. 149 pp. with 206 dot distribution maps, graphs, and 2 colored pl. 34 x 27 cm. [300] Deals with distribution of the world's supply of food and other important agricultural products and with climate, soil, and economic conditions that account for these distributions. Now largely out of date.

GEOGRAPHY OF MINERAL RESOURCES

Mining, periodical indexes: Mudge-Winchell [1], a 17.—Review by T. T. Read of 6 publications on mineral raw materials, *Geog. Rev.* [476], 28 (1938), 505–506. [301]

Jackson, Lucille. *A Guide to Mineral Industries Literature.* State College, Pennsylvania, 1940. 18 pp. (*Pennsylvania State College Bull. Library Studies,* No. 2.) [302] Classified list of selected abstracts, indexes, special ref. books, periods. etc.

———

De Golyer, E., and Harold Vance. *Bibliography on the Petroleum Industry,* 762 pp. Constituting *Bull. of the Agric. and Mechanical College of Texas,* 4th Ser., Vol. 15, No. 11, Sept. 1, 1944; also *School*

of Engineering, Texas Engineering Experiment Station, Bull.,
No. 83. [303]
Topical cl. (system.), incl. a section (areally cl.) on geog. distribution
of petroleum, oil fields, properties, and districts, descriptions and maps;
subj. index; no auth. index.

Atlas

World Atlas of Commercial Geology. (United States Geological Survey.) Washington, 1921. Part 1, "Distribution of Mineral Production," 72 pl., 72 pp. text. Part 2, "Water Power of the World," 8 pl.,
39 pp. text. 35 x 27 cm. [304]
Based for most part on data for year 1913.

LOCATION OF INDUSTRY

McDonald, D. M. *A Select Bibliography on the Location of Industry.*
Montreal, 1937. 95 pp. (*McGill Univ. Social Research Bull.,*
No. 2.) [305]
4 main parts: (1) genl. refs.; (2) industrial location; (3) industrial and
economic surveys; (4) statistical, bibl., and other sources. Auth. index,
index to journals.

Palander, Tord. *Beiträge zur Standortstheorie,* Uppsala, Akademisk
Avhandling, 1935. [306]
Bibl. on pp. 396–416 (topical cl.).

GEOGRAPHY OF TRANSPORTATION
AND COMMUNICATIONS

Transportation, bibliographies and other reference works: Mudge-
Winchell [1]: aeronautics, a 209, b 32, c 43, d 44; railroads and
water transportation, a 160. [307]
Bibliographical data on road transportation: note in *Geog. Rev.*
[476], 27 (1937), 333–334. [308]

*Rivers, Canals and Ports: Bibliographic Notes Giving the List of the
Principal Works Which Have Appeared and of Articles Published
in Periodicals of all Countries.* 6 vols. 1892–1906, 1907–1910, 1911–
1915, 1916–1920, 1921–1925 (5th Series), 1926–1930 (6th Series).
Brussels (Permanent International Association of Navigation Congresses). [309]
Cl. by topics.

Young, Perry. *Bibliographic Notes on Ports and Harbors: Including Lists by the Library of Congress.* New Orleans, American Assn. of Port Authorities, 1926. 188 pp. [310]
2 parts: (1) genl. (topical cl.) and place bibl. (by areas subcl. by ports); (2) subject bibl. (topical cl.).

Pellett, M. E. *Water Transportation: A Bibliography, Guide, and Union Catalogue.* Vol. 1, *Harbors, Ports, and Port Terminals.* New York, Wilson, 1931. 740 pp. [311]
Topical and areal cl.; auth. and subj. indexes.

Atlas

Philip's Mercantile Marine Atlas. . . . Specially designed for use by merchant shippers, exporters and ocean travellers. Ed. by George Philip. 12th ed. London, Geo. Philip, [1926?]. 42 pl. containing over 200 charts and maps. 36 pp. text, statistics, etc. 39 x 53 cm. [312]

Political and Military Geography

(See also [722, 845], periodicals.)

Political science: Mudge-Winchell [1]: encyclopedias, a 114-115; governments, a 115-120, c 24-25.—Selected references to recent works on political geography will be found in *Global Politics,* ed. by R. H. Fitzgibbon, Berkeley, University of California Press, 1944, pp. 183-189 (also in *Journ. of Geog.* [478], December, 1942, pp. 336-340). [313]

Geographical Foundations of National Power. Washington, 1944. (*Army Service Forces Manual* M 103-1, M 103-2, M 103-3.) [314]
Textbook prepared collaboratively for use in Army Specialized Training Program by a group of geographers and others, including Derwent Whittlesey, C. C. Colby, J. K. Wright, Harold Sprout, and Dorothy Good. Contains selective bibls. of works in English dealing with political and economic geog. in genl. and with the political and economic geog. of France, Germany, Italy, British Empire, United States, U.S.S.R., China, Japan, and Latin America. See also [129a].

Foreign Affairs Bibliography: A Selected and Annotated List of Books on International Relations. New York, published by Harper & Bros. for the Council on Foreign Relations. 2 vols.: (1) covering

1919–32 by W. L. Langer and H. F. Armstrong, 1933, 568 pp., and (2) 1932–1942, by R. G. Woolbert, 1945, 726 pp.   [315]
Based on bibls. appearing in *Foreign Affairs,* with omission of items of minor importance and addition of many titles. Some 7,000 entries for 1919–1932; 10,000 for 1932–1942. Vol. for 1932–1942 in 9 parts: (1) genl. international relations; (2) First World War; (3) Second World War; (4) world; (5) Western Hemisphere; (6) Europe; (7) Asia and the Pacific Area; (8) polar regions; (9) Africa. (1)–(4) subcl. topically, (4)–(9) subcl. areally (in part further subcl. by topics); brief anns.; auth. index.

Kiss, George. "Political Geography into Geopolitics: Recent Trends in Germany," *Geog. Rev.* [476], 32 (1942), 632–645.   [316]
Critiques with bibl. refs. in footnotes.

Hartshorne, Richard. "Recent Developments in Political Geography," *Amer. Political Science Rev.,* 29 (1935), 785–804, 943–966.   [317]
Many bibl. refs. in footnotes.

Vogel, Walther. "Politische Geographie und Geopolitik (1909–34)," *Geog. Jhrb.* [6], 49 (1934), 79–304.   [318]
2 main parts: (1) "systematic," cl. by topics, (2) regional; 3,000 entries; ann.; auth. index.

Bowman, Isaiah. *The New World: Problems in Political Geography.* 4th ed. Yonkers, N.Y., and Chicago, World Book Co., 1928. [319]
Bibl. on pp. 747–775 (by regions, with a section on "major problems").

Lewin, Evans. *Select Bibliography of Recent Publications in the Library of the Royal Colonial Institute Illustrating the Relations between Europeans and Coloured Races.* London, 1926. 62 pp. (*Royal Colonial Inst. Bibls.* [330], No. 3.)   [320]

Geological Society of America. (a) *Bibliography of Military Geology and Geography,* prepared under the direction of W. H. Bucher. [New York], 1941. 18 pp. (b) *Ibid., First Supplement,* prepared by Marie Siegrist (geology) and Elizabeth T. Platt (geography), 1943. 11 pp.   [321]
Suggestive lists covering military geology and mapping, military geomorphology.

III. REGIONAL AIDS AND GENERAL GEOGRAPHICAL PERIODICALS

For quick reference to outstanding statistical and geographical publications on different countries the bibliographical notes in the *Statesman's Yearbook,* in the "Geographical Summaries" in *Encyclopædia Britannica World Atlas* [116], and in the several volumes of *Géographie universelle* [52] are recommended. See also [50].
[322]

Mountain Regions

Farquhar, F. P. "The Literature of Mountaineering," *Appalachia,* No. 88 (Dec., 1939), pp. 508–524; No. 89 (June, 1940), pp. 72–95.
[323]
A guide to the outstanding mountaineering books (incl. a few refs. to period. arts.) written in English, with some French, German, and Italian titles. Topical and areal. cl.

Bühler, Hermann. *Alpine Bibliographie.* Munich, Verein der Freunde der Alpenvereinsbucherei. 1931–[1938]. Annual. Publisher varies.
[324]
Continuation of Dreyer [326], 1927. Covers the Alps and other high-mountain regions. Topical and areal cl.; 3,066 entries (1937); genl. index.

Peattie, Roderick. *Mountain Geography: A Critique and Field Study.* Cambridge, Mass., Harvard University Press, 1936. [325]
Refs. to selected books and arts. on the physical and human geog. of mountains and mountain regions follow the chapters and are also presented in an appendix (pp. 245–249) entitled "Bibliographical Notes on the Geomorphology of Mountains."

Dreyer, A. *Bücherverzeichnis der Alpenvereinsbücherei, mit Verfasser- und Bergnamen-Verzeichnis.* (Hauptausschuss des Deutschen und Österreichischen Alpenvereins und Verein der Freunde der Alpenvereinsbücherei.) Munich, J. Lindauerschen Universitäts Buchhandlung, 1927. xvi pp. and 1358 columns (pages not numbered).
[326]

Alph. by regional and mountain names and by other topics relating to mountains and mountaineering; under each hdg. refs. and cross refs. to books and separates from serial publs.; topical index (system.); auth. index; index to mountain names.

A survey by Alois Dreyer of mountaineering clubs throughout the world, with the titles of their organs, will be found in *Alpines Handbuch . . . herausgegeben vom Deutschen und Österreichischen Alpenverein,* Vol. 2, Leipzig, Brockhaus, 1931, pp. 401–422. [327]

Periodical

Revue de géographie alpine
    Institut de Géographie Alpine, University of Grenoble. 1913–. Quarterly. In 1920 the title was changed from *Recueil des travaux de l'Institut de Géographie Alpine.*   [328]
    Devoted to the scientific study of mountain geog. Prior to 1920 devoted more particularly to the special study of the French Alps.

Colonial Empires and Interests

The following bibliographies compiled by L. J. Ragatz were published by Arthur Thomas, London (1 and 2 also by Pearlman, Washington) : (1) *Colonial Studies in the United States During the Twentieth Century,* London, 1932; Washington, 1934, 48 pp. (also in Martineau [331], 1932, 186–237); (2) *A List of Books and Articles on Colonial History and Overseas Expansion Published in the United States in 1931 and 1932,* London, 1934; Washington, 1935, 41 pp.; (3) [*ibid.*] *in 1933, 1934, and 1935,* London, 1936, 97 pp.; (4) *A Bibliography of Articles, Descriptive, Historical, and Scientific, on Colonies and Other Dependent Territories, Appearing in American Geographical and Kindred Journals Through 1934,* London, [1935], 2 vols., 129, 92 pp.   [329]

Royal Empire Society Bibliographies. London, Royal Empire Society, 1915–. Title of Nos. 1–4: *Royal Colonial Institute Bibliographies.*   [330]
Selective bibliographies dealing with the British Empire and the colonial possessions of other nations. The numbers of greatest geog. interest are listed elsewhere in the present volume under the appropriate regional hdgs.

Martineau, Alfred, Paul Rousier, and J. Tramond, eds. *Bibliographie d'histoire coloniale (1900–1930)*. (Premier Congrès International d'Histoire Coloniale, Paris, 1931.) Paris, Société de l'Histoire des Colonies Françaises, 1932. 683 pp.   [331]
Coöperative work on works published 1900–1930 dealing with the colonial hist. of different nations. Each chapter is introduced by running comments in French or English. The cl. of the refs. varies from chapter to chapter. No index. Refs. to the individual chapters will be found in the present work under the appropriate regional hdgs.

Hill, Winifred C. *A Select Bibliography of Publications on Foreign Colonization—German, French, Italian, Dutch, Portuguese, Spanish, and Belgian*. London, 1915. 48 pp. (*Royal Colonial Institute Bibliographies* [330], No. 1.)   [332]
Areal cl.; a few anns.

THE BRITISH EMPIRE

Mudge-Winchell [1]: government documents, a 372–374; historical sources, a 356–357.—In each issue of the *Dominions Office and Colonial Office List* (1862–[1940]; *Colonial Office List,* prior to 1926) will be found a regionally arranged list of parliamentary and nonparliamentary (but official) publications on the affairs of the Dominions and colonies.   [333]
Associations for furthering the interests of geographical education in the British Empire have been established in North Ireland, South Africa, and New Zealand. These associations are affiliated with the Geographical Association, which also maintains branches throughout Great Britain. Information regarding these affiliated associations and branches and their publications may be obtained from the Clerk of the Geographical Association, % Municipal High School of Commerce, Princess Street, Manchester.   [334]

Lewin, Evans. *Best Books on the British Empire: A Bibliographical Guide for Students*. 2d ed. London, 1945. 104 pp. (*Royal Empire Soc. Bibls.* [330], No. 12.)   [335]
Aims to include "the best recent works [books, in English] on the general description, administration, economics, history, and native races of the Dominions, Colonies, and India." Areal cl.; auth. index.

Coupland, R., W. Hill, Evans Lewin, and A. P. Newton. "Grande

Bretagne et Dominions," in Martineau [331], 1932, pp. 413–511.
Running comment (pp. 413–446) followed by "Select List of Publica- [336]
tions, issued since the year 1900, dealing with British Colonial History
(excluding Canada)"; for Canada see [337]. Topical and areal cl.;
most of the refs. are to books; a few to period. arts.

Lewin, Evans. *Subject Catalogue of the Library of the Royal Empire
Society, Formerly Royal Colonial Institute.* London, Royal Empire
Society. Vol. 1, *The British Empire Generally and Africa,* 1930,
856 pp.; Vol. 2, *The Commonwealth of Australia, the Dominion of
New Zealand, the South Pacific, General Voyages and Travels, and
Arctic and Antarctic Regions,* 1931, 770 pp.; Vol. 3, *The Dominion
of Canada and Its Provinces, the Dominion of Newfoundland, the
West Indies and Colonial America,* 1932, 827 pp. [337]
Supplants catalogues published in 1881, 1886, 1895, and 1901, though
"it does not entirely supersede that published in 1901." "Very few publications of any present use or importance have been omitted." "Compiled on a geographical basis, each country being divided under definite
subject heads." Includes refs. not only to books and pamphlets "but also
the titles of articles appearing in the principal reviews and journals,
papers read before learned and other societies and a vast amount of
literature that is not usually included in a catalogue, such, for example,
as analytical contents of numerous volumes of papers and essays." Auth.
index and index to regional and place hdgs. in each vol. The regional
sections cover, in addition to all parts of the British Empire, the South
Pacific, the French Pacific, and the polar regions, in Vol. 1; the whole
of non-British Africa in Vol. 2; and in general the United States and
Central and South America in Vol. 3. Refs. to particular parts of this
great catalogue are given in the notes under the pertinent regional
hdgs. in the present work.

Lewin, Evans. *Select List of Publications in the Library of the Royal
Colonial Institute Illustrating the Communications of the Overseas
British Empire, With Special Reference to Africa Generally and
the Bagdad Railway.* London, 1927. 76 pp. (*Royal Colonial Institute
Bibliographies* [330], No. 4.) [338]
Topical cl., subcl. areally.

Periodical

Bulletin of the Imperial Institute
London. 1903–. Quarterly. [339]

A section devoted to "Recent Progress in Agriculture and the Development of Natural Resources" gives a summary "of the more important
papers and reports received during the preceding quarter, in so far as
these relate to tropical agriculture and the utilisation of the natural
resources of the Colonies, India and the Tropics generally." Also book
reviews on these subjs.

THE FRENCH COLONIAL EMPIRE

Martineau [331], 1932, pp. 238–412.   [340]

Favitski de Probobysz, A. de. *Répertoire bibliographique de la littérature militaire et coloniale française depuis cent ans.* Paris, 1935,
374 pp.   [341]
Lists books in French primarily on military hist., published since 1830
in France, Belgium, Switzerland, and Canada; alph. by auths. and titles;
7,943 entries; no anns.; genl. index.

Atlas

Grandidier, G. *Atlas des colonies françaises, protectorats et territoires
sous mandat de la France.* Paris, Société d'Éditions Géographiques,
Maritimes, et Coloniales, 1934. 39 pl.; text; indexes. 38 x 44
cm.   [342]
Issued in parts for binding in loose-leaf folder; each part deals with a
region and consists of one or more plates, with accompanying text
separately paginated. Locational maps with hypsometric tints; also
geology, rainfall, plant geog., economic geog., etc.

THE NETHERLANDS COLONIAL EMPIRE

Mudge-Winchell [1]: bibliographies, a 358–359.—Martineau [331],
1932, pp. 630–639.   [343]

Hartmann, A., and others. *Repertorium op de literatuur betreffende
de Nederlandsche Koloniën voor zoover zij verspried is in tijdschriften, periodieken, serie—en mengelwerken.* The Hague, Nijhoff, 1895–1937. 9 vols.   [344]
Title varies slightly. The original vol., published in 1895, covers the
Netherlands East Indies for the period 1866–1893 and the Netherlands
West Indies for the period 1840–1894. It is followed by 8 supplements
(*Verfolgen*) (for details see Mudge-Winchell [1], a 359) covering
the period through 1932, and is a continuation of J. C. Hooykaas,

Repertorium op de koloniale litteratuur . . . over de koloniën in mengel-werken en tijdschriften, van 1595 tot 1865 uitgegeven in Nederland en zijne overzeesche bezittingen, Amsterdam, Van Kampen, 1877, 1880. 2 vols.
> By topics (incl. geog., geology, hist., govt., economics, etc.), of which a few are subcl. areally; genl. index.

Kleiweg de Zwaan, J. P. *Anthropologische bibliographie van den Indischen archipel en van Nederlandsch West-Indië,* constituting *Mededeelingen van het Bureau voor de Bestuurzaken der Buitenge-westen bewerkt door het Encyclopaedisch Bureau,* Vol. 30, 1923. 473 pp.   [345]
> 2 separate sections for period. arts. and 2 for books; ann.; genl. index.

Atlas

Atlas van tropisch Nederland. Uitgegeven door het Koninklijk Ne-derlandsch Aardrijkskundig Genootschap in samenwerking met den Topografischen Dienst in Nederlandsch-Indië. 1938. 31 pl., 17 pp. (index). 34 x 43 cm.   [346]
> Pls. 1–10: physical geog., minerals, population, historical geog., etc. of Netherlands Indies as a unit; Pls. 11–31 regional maps, both locational and special.

GERMAN COLONIAL INTERESTS

Comprehensive bibliographical references on the borderlands of Ger-many and on German populations and the regions which they in-habit in other parts of the world, will be found following the apha-betically arranged articles in *Handwörterbuch des Grenz und Aus-landdeutschtums,* edited by Carl Petersen and others, Breslau, Hirt, 1933–. (Was to be completed in 5 vols.; had reached the letter "L" in 1939; see *Geog. Rev.* [476], 27 (1937), 689–690.)   [347]

Periodicals

Koloniale Rundschau: Zeitschrift für koloniale Länder, Völker und Staatenkunde
> Berlin, Leipzig. 1909–[1940]. Subtitle varies. Combined with *Mitteilungen aus den Deutschen Schutzgebieten* [350] during 1929–1930.   [348]

Deutschtum im Ausland: Zeitschrift des Deutschen Ausland-Instituts

Stuttgart. 1918–1940. Fortnightly. Title prior to 1934, *Der Aus-landdeutsche.*   [349]
Brief but good lists of current books, maps, and arts. dealing more especially with the interests of Germans abroad and also with matters of general and economic interest.

Mitteilungen aus den deutschen Schutzgebieten
    Berlin, E. S. Mittler & Sohn. 1888–[1928/9]. Irregularly. Combined with *Koloniale Rundschau*, 1929–30. Indexes: 1888–1897; 1898–1928/9 (*Ergänzungsheft* 16, 1930).   [350]
Arts. and maps on political, economic, and geog. problems of the former German colonies. *Ergänzungshefte*, comprising monographs, published irregularly, 1908–1930.

Deutsche Erde: Zeitschrift für Deutschkunde: Beiträge zur Kenntnis Deutschen Volkstums Allerorten und Allerzeiten
    Gotha, Justus Perthes. 1902–1915. 8 numbers a year.   [351]
Contains extensive bibl. notes on German peoples and folklore throughout the world.

SPANISH COLONIAL INTERESTS

Spanish colonial history, 1900–1930: Martineau [331], 1932, pp. 121–185.   [352]

Hilton, Ronald, ed. *Handbook of Hispanic Source Materials and Research Organizations in the United States*. Toronto, University of Toronto Press, 1942. 455 pp.   [353]
"This Handbook consists of a series of statements describing collections of Hispanic source materials and organizations engaged in Hispanic research in the United States. The word 'Hispanic' is used in its widest sense; it comprises Spain, Portugal and Hispanic America of the pre- and post-Columbian periods; Florida, Texas, and Southwest, and California are included until their annexation by the United States. The material surveyed belongs to the fine arts, the humanities, and the social sciences; the natural sciences have been included in the case of some exceptional collections." Alph. by states, subcl. by localities, further subcl. by institutions; genl. index.

Spain and Spanish America in the Libraries of the University of California: A Catalogue of Books. Vol. 1, *The General and Departmental Libraries*, Berkeley, Cal., 1928, 852 pp.; Vol. 2, *The Bancroft Library*, 1930, 844 pp.   [354]

"The entries cover books in the Spanish language or relating to Spain and Spanish America, including those on Spanish exploration, colonization, and rule in other countries." Each vol. is cl. by auths., and each has an index of topics and geog. names (system. in Vol. 1; alph. in Vol. 2).

The Polar Regions

Lewin [337], Vol. 2, 1931, pp. 652–664.—Norwegian possessions and spheres of influence: Martineau [331], 1932, pp. 614–629.   [355]

Dutilly, Artheme. *Bibliography of Bibliographies on the Arctic.* Washington, D.C., The Catholic University of America, 1945. 47 pp.   [355a]
Alph. by auths. and titles.

Platt, R. R. "Recent Exploration in the Polar Regions," *Geog. Rev.* [476], 29 (1939), 303–309.   [356]
Review article on recent publs.

U.S. Works Progress Administration, New York City. Bibliographies and Indices of Special Subjects Project No. 465-97-3-18. *Selected List of Bibliographies on the Polar Regions.* New York, [1938]. Part 1, 41 pp.; Part 2, 27 pp. (*Annotated Bibliography of the Polar Regions,* Ser. B.)   [357]
Covers not only independent bibls. but also incidental bibls. in books, period. arts., etc.; by auths.

Bistrup, H. A. Ø., ed. *Katalog over litteratur vedrørende polaromraadernes og verdenshavenes, opdagelse og udforskning, hval-og saelfangst, personalhistorie, tidsskrifter, aarsskrifter og andre periodica. Catalogue of Literature Concerning the Discoveries and the Explorations of the Polar Environs and the Oceans, Whale- and Sealfisheries, Biography, Periodicals, Annuals and Other Periodical Papers.* (Denmark, Marinens Bibliothek: The Library of the Danish Navy.) Copenhagen, 1933. 406 pp.   [358]
By topics and geog. names subcl. by topics; list of periods., genl. index.

Hulth, J. M. *Swedish Arctic and Antarctic Explorations 1758–1910: Bibliography.* Uppsala and Stockholm, 1910. 189 pp. (*Kongliga Svenska Vetenskapsakademiens Årsbok för år 1910,* Bilaga 2.)   [359]
Topical cl.; auth. index.

Chavanne, Josef, Alois Karpf, and Franz Ritter von Le Monnier. *The Literature on the Polar Regions of the Earth. Die Literatur über die Polar-Regionen der Erde.* (K.K. Geographische Gesellschaft in Wien.) Vienna, 1878. 346 pp. Title and preface in English and German.   [360]
Place cl., subcl. by topics, with subhdgs. for books, period. arts., and maps; 6,617 entries; auth. index.

Periodicals

The Polar Record: Printed in Great Britain for the Scott Polar Research Institute
Cambridge, England, University Press. 1931–. Semiannual. [361]
Lists current polar literature.

Polar-Årboken
Oslo, Norsk Polarklubb. 1933–[1937]. Annual.   [362]

THE ARCTIC REGIONS

(See also [238a], Arctic Sea.)

Arctic soils: Giesecke [230], 1939, pp. 285–286.   [363]

Breitfuss, Leonid. "Das Nordpolargebiet (1913–31)," *Geog. Jhrb.* [6], 47 (1932), 129–270.   [364]
Section on north polar regions in genl. subcl. by topics; sections on Asiatic and Greenland-American sectors, each subcl. areally, further subcl. by topics; ann.; 1,826 entries; auth. index.

Breitfuss, Leonid. "Die Erforschung des Nordpolargebiets in den Jahren 1913–28: Die innere Arktis und der europäische Sektor," *Geog. Jhrb.* [6], 44 (1929), 289–374.   [365]
By topics and areas, subcl. by topics; ann.; 1,276 entries; auth. index.

Greenland: Adams and Norwell [442], 1928–1936.   [366]

Union Catalogue of Literature on Greenland. Chicago, 1940. 49 mimeo. pp. (*The John Crerar Library Reference List No. 45.*)   [366a]
Product of a W.P.A. project. Covers American contributions to the exploration of Greenland, with indication of holdings of four libraries in Chicago. Books and period. arts. are listed separately (alph. by auths.).

Vartdal, Hroar. *Bibliographie des ouvrages norvégiens relatifs au Groenland (y compris les ouvrages islandais antérieurs à l'an 1814)*, constituting *Skrifter om Svalbard og Ishavet* [373], Vol. 54. Oslo, J. Dybwad, 1935. 119 pp. [367]
By topics, incl. history, expeditions, the natural sciences, economics, languages; auth. index.

Lauridsen, P. *Bibliographia groenlandica: eller Fortegnelse paa Vaerker, Afhandlinger og danske Manuskripter, der handle om Grønland indtil Aaret 1880 incl.*, constituting *Meddelelser om Grønland* [372], Vol. 13, 1890. 247 pp. [368]
By topics, incl. voyages, geog., the natural sciences, history, economic and social life, Eskimos, maps; auth. index.

Taracouzio, T. A. *Soviets in the Arctic: An Historical, Economic, and Political Study of the Soviet Advance into the Arctic.* New York, Macmillan, 1938. [369]
Bibl. on pp. 503–546; sections on documents, treatises, arts., periodicals.

Arctic Institute of U.S.S.R. *A Bibliography of Novaya Zemlya.* (The Chief Administration of the Northern Sea Route.) Leningrad, 1935. 240 pp. Also has Russian title. [370]
"The first issue of a series of bibliographical guides . . . which the Arctic Institute intends to publish." By topics, incl. bibls., maps, hist., expeditions, the natural sciences; anns. in Russian; 2,645 entries.

"Swedish Explorations in Spitzbergen, 1758–1908," *Ymer* [647], 1909, pp. 1–89 (also reprinted separately, Stockholm, 1909. 89 pp.). [371]
Includes: J. M. Hulth, "Swedish Spitzbergen Bibliography," pp. 23–77 (376 entries), and Gerard De Geer, "Swedish Spitzbergen Maps Until the End of 1908," pp. 78–89 (60 entries), both by topics.

Periodicals

Meddelelser om Grønland

Copenhagen, Kommissionen for Videnskabelige Undersøgelser i Grønland. 1876–. Irregularly. Prior to 1931 published by Kommissionen for Ledelsen af de Geologiske og Geografiske Undersøgelser i Grønland. Index: 1876–1926. [372]
Arts. on Greenland history, ethnography, geology, and geog. in several languages.

Skrifter om Svalbard og Ishavet
 Oslo, Norges Svalbard og Ishavs-Undersøkelser. 1929–. Irregularly. The same institution has also issued *Meddelelser* since 1926.   [373]

Sovetskaia Arktika: Ezhemesiachnyĭ politiko-ékonomicheskiĭ zhurnal [Soviet Arctic: Monthly Politico-Economic Journal]
 Moscow, Chief Administration of the Northern Sea Route. 1935–[1940]. Monthly.   [374]
 In Russian.

Problemy Arktiki: Problems of the Arctic
 Leningrad, The Arctic Institute. 1937–[1940]. Monthly.   [375]
 Arts. (a few with English résumés) and bibls. In Russian.

Arctica
 Leningrad, The Arctic Institute. 1933–[1937]. Annual.   [376]
 Arts. in Russian, German, English; résumés in German or English of Russian arts. (See *Geog. Rev.* [476], 24 (1934), 499.)

Biulleten' Arktĭcheskogo Ĭnstĭtuta SSSR: Bulletin of the Arctic Institute, U.S.S.R.
 Leningrad. 1931–[1936]. Weekly.   [377]
 In Russian. English résumés.

THE ANTARCTIC REGIONS

Oceanography of Antarctic: Schulz [236], 1939, pp. 83–84.   [378]

Breitfuss, Leonid. "Das Südpolargebiet (1913–32)," *Geog. Jhrb.* [6], 48 (1933), 101–158.   [379]
 By topics, incl. expeditions and the natural sciences; 661 entries; auth. index.

Aagaard, Bjarne. *Fangst og forskning i Sydishavet.* Oslo, Gyldendal Norsk Forlag, 1930. 2 vols.   [380]
 Two bibls. in Vol. 2, one on the Antarctic (pp. 957–1032) and one on whaling (pp. 1033–1068), both by auths.

Denucé, Jean. "Bibliographie antarctique," in Commission Polaire Internationale, *Procès-verbal de la session tenue à Rome en 1913 présenté par G. Lecointe,* Brussels, 1913, pp. 25–293.   [381]

By topics, incl. bibls., maps, the natural sciences, economics, expeditions; 3,225 entries; auth. index.

Mill, H. R. "A Bibliography of Antarctic Exploration and Research," in *The Antarctic Manual,* published by the Royal Geographical Society, London, 1901, pp. 517–586.   [382]
Chron.; 878 entries; index of auths. and explorers.

———

Sparn, Enrique. "Segunda contribución al conocimiento de la bibliografía meteorológica y climatológica del quadrante americano de la Antártica y Subantártica," *Boletín de la Academia Nacional de Ciencias,* Córdoba, Argentina, 34, Pt. 2a (1938), 183–201.   [383]
Auth. cl.; topical index (system.). Supplements E. Sparn, "Literatura sobre meteorología y climatología de las regiones antártica y subantártica americanas," in *Acad. Nac. de Ciencias, Miscelánea,* No. 7, Córdoba, 1923, pp. 57–72; see [600].

Atlas

Hansen, H. E., ed. *Atlas over Antarktis og Sydishavet.* Utgitt av Hvalfangernes Assuranceforening. [Oslo, 1936?]. 15 pl.; 4 large maps in pocket. 34 x 41 cm.   [384]
Explorers' routes, pack ice, wind and temperatures, geology, plant geog., etc.

The Atlantic Ocean and Its Islands

Oceanography of Atlantic Ocean: Schulz [236], 1939, pp. 51–83.   [385]

ICELAND

Hermannsson, Halldór. *Catalogue of the Icelandic Collection Bequeathed by Willard Fiske.* (Cornell University Library.) Ithaca, N.Y., Cornell Univ. Press, 1914. 763 pp. *Additions 1913–26,* 1927, 284 pp.; *1927–1942,* 1943, 295 pp.   [386]
Covers one of the largest Icelandic collections in the world and deals to some extent with Scandinavian publications in genl. By auths.; subj. indexes.

Iwan, Walter. *Island: Studien zu einer Landeskunde.* Stuttgart, J. Engelhorn, 1935. 155 pp. (*Berliner geographische Arbeiten,* No. 7.)   [387]
Bibl. on pp. 129–150: by auths.; 623 entries.

Klose, Olaf. *Islandkatalog der Universitätsbibliothek Kiel und der Universitäts- und Stadtbibliothek Köln.* Kiel, Universitätsbibliothek, 1931. 423 pp. (*Kataloge der Universitätsbibliothek Kiel, 1.*)   [388]

Nielsen, Niels. "Island (1914–30)," *Geog. Jhrb.* [6], 46 (1931), 219–226.   [389]
Topical cl.; 114 entries; ann.; auth. index.

Series

Islandica: An Annual Relating to Iceland and the Fiske Icelandic Collection in Cornell University Library. Ithaca, N.Y., Cornell Univ., 1908–. Irregularly.   [390]
This series has included much bibl. material. Numbers have been devoted to Icelandic literature, sagas, eddas, manuscripts, and periodicals.

OTHER ISLANDS

Falkland Islands: Lewin [337], Vol. 2, 1931, pp. 571–573.   [391]

Tittelbach, Gertrud. *Beiträge zur Landschaftskunde von Teneriffa.* Hamburg, 1931.   [392]
This doctoral dissertation (Univ. of Hamburg) contains a bibl. (pp. 98–103) by titles (chron.); 88 entries.

Hartnack, Wilhelm. *Madeira: Landeskunde einer Insel.* Hamburg, Friederichsen, de Gruyter, 1930.   [393]
Bibl. (pp. 154–177): by topics, incl. maps; 423 entries.

Schütz, J. F. *Bausteine zu einer Bibliographie der Canarischen, Madeirischen, und Capverdischen Inseln und der Azoren (bis einschl. 1920).* Graz, U. Moser, 1929. 144 pp.   [394]
Auth. cl.; 1,350 entries.

Cole, G. W. *Bermuda in Periodical Literature, with Occasional References to Other Works: A Bibliography.* Boston, 1907. 285 pp.   [395]
By auths. of books and publishers or titles of periods. (alph.); ann.; genl. index.

The Pacific Ocean and Region

(See also [186], early cartography; [269], medical geography; [1140a], best books; [1173], periodical.)

Mudge-Winchell [1]: gazetteer, a 326.—Lewin [337], Vol. 2, 1931, pp. 483–570.—Oceanography: Schulz [236], 1939, pp. 41–49.— Colonial history, 1900–1930: Martineau [331], 1932, pp. 395–405.— Economic geography: Lütgens [284], 1936, pp. 123–125.   [396]

References on the Physical Oceanography of the Western Pacific Ocean. (United States Navy Department: Hydrographic Office.) Washington, 1946. 183 pp. (*H.O. Pub.*, No. 238.)   [396a] Compiled by Mary C. Grier. Alph. by auths.; 1,227 entries; topical index; index to geog. names and expeditions.

Conover, Helen F. *Islands of the Pacific: Supplement.* (The Library of Congress, General Reference and Bibliography Division.) Washington, 1945. 68 mimeo. pp.   [397] 389 entries; auth. and subj. indexes covering both the 1943 edition [400] and this supplement.

Allied Geographical Section: Southwest Pacific Area. *An Annotated Bibliography of the Southwest Pacific and Adjacent Areas.* To be completed in 4 vols.; Vols. 1–3, n.p., 1944; Vol. 4, forthcoming. Vol. 1, *The Netherlands and British East Indies and the Philippine Islands,* 317 pp.; Vol. 2, *The Mandated Territory of New Guinea, Papua, the British Solomon Islands, the New Hebrides, and Micronesia,* 274 pp.; Vol. 3, *Malaya, Thailand, Indo China, the China Coast, and the Japanese Empire,* 256 pp. Vol. 4, Supplement. [398] "A selection of geographical literature available in Australia considered to be of use for military purposes," compiled with the collaboration of 70 Australian libraries and societies. Areal cl.; the ann. for each work indicates its contents (incl. maps. charts, and photographs) and the libraries in Australia in which it may be found.

Lewin, Evans. *The Pacific Region: A Bibliography of the Pacific and East Indian Islands, Exclusive of Japan.* London, 1944. 76 pp. (*Royal Empire Society Bibliographies* [330], No. 11.)   [399] By places and topics, incl. bibls. (both independent bibls. and incidental bibls. in books); auth. and place indexes.

Conover, Helen F. *Islands of the Pacific: A Selected List of References.* (The Library of Congress, Division of Bibliography.) Washington, 1943. 181 mimeo. pp.   [400] By topics (incl. bibls.) and places; 1,747 entries; auth. index; subj. index (topics subcl. by places; places subcl. by topics). See also [397].

Peake, C. H. "War and Peace in the Pacific: A Classified and Anno-

tated Bibliography of Selected Books," *Far Eastern Quarterly*, 1, (1943), 253–276.   [400a]
"Lists 169 of the better and more general introductory books relating to the area," and also "the leading published bibls. in the field."

Ellinger, W. B., and Herbert Rosinski. *Sea Power in the Pacific, 1936–1941: A Selected Bibliography of Books, Periodical Articles, and Maps from the End of the London Naval Conference to the Beginning of the War in the Pacific.* Princeton, Princeton University Press; London, Humphrey Milford, Oxford University Press, 1942. 94 pp.   [401]
By areas and topics (incl. naval geog. of the Pacific, pp. 7–12), subcl. by books, arts., bibls., and maps; 600 entries; ann.; auth. index.

Wright, J. K. "Pacific Islands," *Geog. Rev.* [476], 32 (1942), 481–486.   [402]
Review article on recent publications.

Grier, Mary C. *Oceanography of the North Pacific Ocean, Bering Sea and Bering Strait: A Contribution Toward a Bibliography.* Seattle, University of Washington, 1941. 312 pp. (*University of Washington Publs., Library Series*, Vol. 2, May, 1941.)   [403]
By topics, incl. physical oceanography (53 pp.) and marine biology (195 pp.); 2,929 entries; a few anns.; auth. index; index of topics, geog. names, and titles of books.

Merrill, E. D. *Polynesian Botanical Bibliography 1773–1935.* Honolulu, 1937. 194 pp. (*Bernice P. Bishop Museum Bull.*, No. 144.)   [404]
Auth. cl.; ann.

——

Reid, C. F., ed. *Bibliography of the Island of Guam.* New York, H. W. Wilson, 1939. 102 pp.   [405]
Topical cl. (alph.); ann.

Gusinde, Martin. "Bibliografia de la Isla de Pascua," *Publicaciones del Museo de Etnología y Antropología de Chile,* Santiago de Chile, 2, No. 2 (1920), 201–260; 2, No. 3 (1922), 261–383.   [406]
Genl. bibl. discussions of a variety of topics, each followed by refs.; also a bibl. of the anthropology, ethnology, and language of the island.

——

Periodicals

L'Océanie française: Bulletin du Comité de l'Océanie Française

Paris. 1904–[1940]. Monthly. [407]
Corresponds to *L'Afrique française* [1057].

Bulletin de la Société des Études Océaniennes
Papeete, Tahiti. 1917–. Irregularly. [408]

Maps

MacFadden, C. H. *A Bibliography of Pacific Area Maps*. San Francisco, New York, etc., 1941. 130 pp. (American Council, Institute of Pacific Relations, *Studies of the Pacific*, No. 6.) [409]
By areas, incl. "World," "Pacific Ocean," "Oceania," and the continents adjoining the Pacific; 290 entries; ann.; subj. index. For each map, holdings in certain collections in Washington, New York, Syracuse, and Ann Arbor are indicated.

THE AMERICAS

Political geography: Vogel [318], 1934, pp. 293–303.—The section "America" in Lewin [337], Vol. 3, 1932, pp. 634–691, includes references on cartography and, under the headings "Discovery and Exploration" and "History," to works dealing with specific regions (topical and place cl. (system.); auth. index). [410]

Griffin, Grace G., and others. *Writings on American History* [1906–1938]: *A Bibliography of Books and Articles on United States and Canadian History for the Year . . . with Some Memoranda on Other Portions of America*. Publisher varies. 1908–[1942]. (Published as part of *Ann. Rep. of the Amer. Hist. Assn.* Washington, Govt. Print. Off., 1918–[1942].) Normally annual. [411]
The vols. for 1906–1932 appeared under the sole authorship of Grace G. Griffin; others have collaborated with her in the subsequent vols.
The two more recent vols. covering 1936 and 1937–1938 deal exclusively with United States history except for relatively short sections on "generalities" and "America in general." Topical cl. (system., alph., and chron.); a section "Regional History" subcl. by states; 6,722 entries (1937–38); genl. index.

Phelps, Elizabeth, ed. *Statistical Activities of the American Nations, 1940: A Compendium of the Statistical Services and Activities in 22 Nations of the Western Hemisphere, Together with Information Concerning Statistical Personnel in These Nations.* Washington, Inter-American Statistical Institute, 1941. 873 pp. [412]

Aims "to disseminate knowledge of the statistical activities related to the collection, processing, and publication of social and economic information"; many bibl. data; cl. by countries.

Sabin, Joseph, and others. *Bibliotheca Americana: A Dictionary of Books Relating to America, from Its Discovery to the Present Time.* New York, Published by the New York Public Library for the Bibliographical Society of America, 1868–1936. 29 vols. [413] Title varies. Begun by Joseph Sabin, continued by Wilberforce Eames, and completed by R. W. G. Vail. This monumental work lists by auth. a wide selection of books, pamphlets, and periods. printed in or relating to America and indicates libraries where copies of rare books may be found, but in genl. is of value only for works of which the author's name is known.

Eberhardt, Fritz. *Amerika-Literatur: die wichtigsten seit 1900 in deutscher Sprache erschienenen Werke über Amerika.* Leipzig, 1926. 335 pp. (*Koehler & Volkmars Literaturführer,* Vol. 7.) [414] Deals with German books on a wide variety of topics incl. geog., hist., ethnography, travel, and science. The American student will find the material on Latin America of particular value. System. by topics, subcl. by places; 1,635 entries; ann.; auth. and subj. indexes.

Work, M. N. *A Bibliography of the Negro in Africa and America.* New York, H. W. Wilson, 1928. 719 pp. [415] Part II, "The Negro in America," pp. 251–660: 3 sections: (1) the Negro in the settlement of America (pp. 251–279); (2) the Negro in the United States (pp. 281–636); (3) present conditions of the Negro in the West Indies and Latin America (pp. 636–660); each subcl. in detail by topics; ann.; auth. index. See also [1044].

————

Fisher, J. E. (a) *Bibliography of United States Mountain Ascents (except Pacific Coast States and Eastern Seaboard),* 64 pp. (b) *Bibliography of Eastern Seaboard Mountain Ascents,* 16 pp. (c) *Bibliography of Alaskan and Pacific Coast Mountain Ascents,* 85 pp. (d) *Bibliography of Canadian Mountain Ascents,* 103 pp. (e) *Bibliography of Mexican, Central and South American Mountain Ascents,* 25 pp. Neither place nor publisher indicated. All dated June 30, 1945. [416] Each vol. contains (1) alph. list by names of mountains, giving for each mountain refs. to arts. and notes in mountaineering journals, and (2) selected lists of libraries containing the journals cited.

Periodicals

Bulletin of the Pan American Union
Washington. 1893–. Monthly. From 1893 to Sept., 1910, the title
was (*Monthly*) *Bulletin of the* (*International*) *Bureau of the
American Republics.* [417]

*Acta Americana: Review of the Inter-American Society of Anthro-
pology and Geography*
Washington, 1943–. Quarterly. Subtitle also in Spanish and Portuguese. [418]
Articles, abstracts, bibl. notes, etc., in several languages.

Journal de la Société des Américanistes de Paris
1895–96–[1940]. Semiannual since 1930. [419]
A scholarly "Bibliographie Américaniste" (topical cl., subcl. geog.),
beginning 1914, appeared in this, covering especially the ethnography
and archeology but also the linguistics, hist., and geog. of America as
a whole. Book reviews by qualified scholars are also included.

North America

(See also [173], voyages of exploration; [184, 185], early cartography; [1160].)

Geomorphology: Spreitzer [214], 1938, pp. 248–251.—Soils: Giesecke
[230], 1939, pp. 274–282. [420]

*A.A.A. Travel Reading: A Selected Bibliography Covering the
United States, Alaska, and Mexico.* Washington, Amer. Automobile
Assn., 1945. 54 pp. Pocket size. [421]
For "the motorist who wishes to obtain the greatest amount of pleasure
and benefit from his travels." By states, with sections on "general background literature" (incl. cultural and social heritages, nature and rec-
reation, waterways), major regions of U.S.A., Alaska, and Mexico; ann.

Murdock, G. P. *Ethnographic Bibliography of North America.* New
Haven, 1941. 168 pp. (*Yale Anthropological Studies,* Vol. i.) [422]
Covers ethnography of Eskimos and Indians. By regions, each region
subcl. by tribes; no indexes.

Joerg, W. L. G. "The Geography of North America: A History of
Its Regional Exposition," *Geog. Rev.* [476], 26 (1936), 640–
663. [423]

Contains hist. and bibl. data (titles (chron.) cl. by countries of publ.) on books dealing with the regional geog. of North America and of its larger parts.

Dietrich, Bruno. "Nordamerika (1916–30)," *Geog. Jhrb.* [6], 45 (1930), 243–390; 46 (1931), 227–340.   [424]
While dealing primarily with the United States and Canada, includes refs. to works on the Americas as a whole. System. by topics and by regions, subcl. by topics; regional bibl. of the U.S. by states (in Vol. 46); 4,533 entries; ann.; auth. index.—See also [1160].

___

Flint, R. F. *Glacial Map of North America: Part 2, Bibliography and Explanatory Notes.* [New York], 1945. 45 pp. (*Geological Soc. of Amer. Special Papers*, No. 60.)   [425]
The bibl. (pp. 7–37) is designed "to provide a well-rounded list of those references judged to be most helpful and significant to a more detailed investigation of the glacial geology of any region" in North America. Alph. by states, provinces, and territories.

U.S. Geological Survey. [Bibliographies of North American geology.]   [426]
The fundamental source for bibl. data on the geology of North America since 1785 is a series of bibls. by J. M. Nickles and Emma M. Thom, constituting *U.S. Geol. Survey Bulls.* 746, 747, and 823 (for period 1785–1928), 937 (for 1929–1939), 938 (for 1940–1941), 949 (for 1942–1943). By auths.; subj. index with topics subcl. areally, and areas subcl. by topics. For further details see Mudge-Winchell [1], a 182–183, b 29, c 36, d 34.

Munns, E. N. *A Selected Bibliography of North American Forestry.* Washington, 1940. 2 vols. (U.S. Dept. of Agriculture, *Misc. Publ.,* No. 364.)   [427]
Refs. to works on forestry published in Canada, Mexico, and the United States prior to 1930. Topical cl. (system.); 21,413 entries; auth. index.

Greene, Amy B., and F. A. Gould. *Handbook-Bibliography on Foreign Language Groups in the United States and Canada.* New York, Council of Women for Home Missions and Missionary Education Movement, 1925. 160 pp.   [428]
General bibls. on topics relating to immigration into the United States and Canada and the assimilation and education of the immigrant, together with special bibls. on 38 individual foreign language groups.

The latter contain refs. to publs. relating to the life of each group in its native country. By topics and by language groups subcl. by topics.

Griffin, A. P. C. *Bibliography of American Historical Societies (The United States and the Dominion of Canada)*. 2d ed. Washington, 1907. 1,374 pp. (*Annual Rep. of the Amer. Hist. Assn. for the Year 1905*, Vol. 2.) [429]
The publs. of the hist. (and also geog.) societies of the United States and of Canada are listed in full. Sections on national assns. (alph.) and on state, provincial, and local historical socs. (alph.); 7,537 entries; genl. index; "personal subject index"; index of socs.

———

Bibliographies of the Pacific Northwest covering 1941 and subsequent years: *Pacific Northwest Quarterly*, 1942–. [430]

Inland Empire Council of Teachers of English. *Northwest Books: Report of the Committee on Books of the Inland Empire Council of Teachers of English, 1942.* . . . Portland, Ore., Binfords & Mort, 1942. 356 pp. [431]
"Chiefly a reference volume of Northwest authors, not necessarily of Northwest writing," the "Northwest" being the states of Idaho, Montana, Oregon, and Washington. Covers both fiction and nonfiction. Part I, "Northwest Books" (auth. cl.; indexes: by titles, by titles cl. by states, by auths. cl. by states). Part II, (1) "Northwest Magazine Bibliography" (alph. by auths. and titles); (2) "Selected, Descriptive List of Books about the Pacific Northwest by Non-Native or Non-Resident Authors" (topical cl.; ann.); (3) lists of public and private libraries.

Appleton, J. B. *The Pacific Northwest: A Selected Bibliography Covering Completed Research in the Natural Resource and Socio-Economic Fields, an Annotated List of In-Progress and Contemplated Research, Together with Critical Comments Thereon, 1930–1939*. Portland, Northwest Regional Council, 1939. 474 pp. [432]
Deals primarily with the northwestern United States; a relatively few refs. to works on Alaska and northwestern Canada. By topics, incl. bibls., maps; 4,635 entries.

Schmieder, Oscar. *Länderkunde Mittelamerikas: Westindien, Mexico und Zentralamerika.* (Separate vol. in Oskar Kende, ed., *Enzyklopädie der Erdkunde.*) Leipzig und Vienna, Franz Deuticke, 1934. [433]
Bibl. on pp. 158–171 (areal cl.; 333 entries).

ALASKA

Mudge-Winchell [1]: handbook, d 24.—Dietrich, "Alaska," *Geog. Jhrb.* [6], 46 (1931), 332–340 (part of [425]).—Adams and Norwell [442], 1928–1936.   [434]

Fuller, Grace H. *Alaska: A Selected List of Recent References.* (The Library of Congress, Division of Bibliography.) Washington, 1943. 181 mimeo. pp.   [435]
Topical cl. (system.); 1,588 entries; auth. index.

Wickersham, James. *A Bibliography of Alaskan Literature, 1724–1924, Containing the Titles of All Histories, Travels, Voyages, Newspapers, Periodicals, Public Documents, etc., Printed in English, Russian, German, French, Spanish, etc., Relating to, Descriptive of, or Published in Russian America or Alaska from 1724 to and Including 1924.* Cordova, Alaska, 1927. 662 pp. (*Alaska Agric. College and School of Mines, Miscellaneous Publs.,* Vol. 1.)   [436]
2 sections: (1) "General Publications Relating to Alaska" (by topics, alph.), and (2) "U.S. Public Documents Relating to Alaska" (by departments); 10,380 entries; genl. index.

Dall, W. H., and Marcus Baker. "Partial List of Charts, Maps, and Publications Relating to Alaska and the Adjacent Regions from Puget Sound and Hakodadi to the Arctic Ocean, Between the Rocky and the Stanovoi Mountains," in: U.S. Coast and Geodetic Survey, *Pacific Coast Pilot: Coasts and Islands of Alaska,* Ser. 2, Washington, 1879, pp. 163–374.   [437]
Includes refs. to Russian materials and to periodical series. Maps cl. by places (alph.); books and other publs. cl. by auths.

———

Fuller, Grace H. *Aleutian Islands: A List of References.* (The Library of Congress, Division of Bibliography.) Washington, 1943. 41 mimeo. pp.   [438]
Sections on (1) bibls. (alph.); (2) books, etc. (alph.); (3) period. arts. (chron.); auth. index.

CANADA AND NEWFOUNDLAND

Mudge-Winchell [1]: Bibliographies, historical, b 49, c 69; bibliographies, national and trade, a 391, c 77; gazetteers and geographical names, a 324, 328; government documents, a 372–373; c 74; statistical yearbooks, a 124.—Lewin [337], Vol. 3, 1932, pp. 1–503.—Martineau [331], 1932, pp. 511–557.—An annual survey of Canadian government publications appears in *The Canada Yearbook* (Dominion Bureau of Statistics, Ottawa).—See also [1170]. [439]

McGill University Library School. *A Bibliography of Canadian Bibliographies.* Compiled by the 1929 and 1930 Classes in Bibliography . . . under the Direction of M. V. Higgins. Montreal, 1930. 45 pp. [440]
Does not include some refs. which one might expect to find. Topical cl. (alph.); ann.; genl. index.

———

Bibliographies on Canadian economics will be found in *University of Toronto: History and Economics: Contributions to Canadian Economics,* as follows: for 1920–1924, in Vol. 3, 1931; for 1925–1927 and 1928–1929, in Vol. 2, 1929; for 1927–1928, in Vol. 1, 1928; for 1930–1934, in Vols. 4–7, 1932–1934. Since 1934 these have been continued in each issue of the *Canadian Journal of Economics and Political Science* under the title "A Bibliography of Current Publications on Canadian Economics."—Graduate theses in Canadian history and economics have been listed in the *Canadian Historical Review* [446], 1927–. [441]

Adams, J., and M. H. Norwell. "A Bibliography of Canadian Plant Geography . . . ," *Trans. of the Royal Canadian Inst.,* 16, Pt. 2 (1928), 293–355; 17, Pt. 1 (1929), 103–145; 17, Pt. 2 (1930), 227–295; 18, Pt. 2 (1932), 343–373; 21, Pt. 1 (1936), 95–134. The parts issued 1928–1930 by J. Adams only. [442]
Includes material on Canada, Greenland, Newfoundland, Labrador, and Alaska. By chron. periods subcl. by auths; ca. 3,500 entries.

Trotter, R. G. *Canadian History: A Syllabus and Guide to Reading.* New York, Macmillan, 1926. 175 pp. [443]
Although primarily for college students of history, there are sections of value to the student of geog. on such subjects as: geog. and descriptive works and atlases, official publs., bibls.; also selected readings (i.e., specific refs. to chapters and pages) in works on the country and its inhabitants, discovery, and exploration, settlement, boundaries, immigration, etc. By forms of publ. and by chron. periods subcl. by topics and places; auth. index.

Gagnon, Philéas. *Essai de bibliographie canadienne: Inventaire d'une bibliothèque comprenant imprimés, manuscrits, estampes, etc., relatifs à l'histoire du Canada et des pays adjacents, avec des notes bibliographiques.* [Vol. 1], Quebec, 1895. 721 pp. Vol. 2, Montreal, 1913. 475 pp. [444]
Although this is a catalogue of a private library, now turned over to the city of Montreal, it is a fundamental record of Canadiana. Alph. by auths. and topics; 10,031 entries.

Perret, Robert. *La Géographie de Terre-Neuve.* Paris, Guilmoto, 1913. [445]
Bibl. on pp. 353–370; titles (chron.); 423 entries.

Periodicals

The Canadian Historical Review
University of Toronto Press. 1920–. Quarterly. Succeeds *Review of Historical Publications Relating to Canada,* annual, published 1896–1918 by the University of Toronto. Indexes: 1896–1905, 1906–1915, 1920–1929, 1930–1939. [446]
Book reviews, review arts., and bibl. notes for the most part on Canadian hist. Attention is also given to travels and geog.

The Canadian Geographical Journal
Ottawa, The Canadian Geographical Society. 1930–. Monthly. Index: 1930–33. [447]
Popular, illustrated magazine.

Bulletin de la Société de Géographie de Québec
1880–1897. New series, 1908–1934. Irregularly. The old series was published in both French and English with the additional title *Transactions of the Geographical Society of Quebec.* Indexes: 1880–1912, 1912–1916, 1916–1918, 1919–1921. [448]
Succeeded 1942–1944 by *Bulletin des Sociétés de Géographie de Québec et de Montréal.*

Maps and Atlases

Official maps: Thiele [97], 1938, pp. 221–231. [449]

Atlas of Canada. Rev. and enlarged ed. Prepared under the direction of J. E. Chalifour, Chief Geographer. (Department of the Interior, Canada.) 1915. 124 pp. maps. 14 pp. text (contents, statistics). 32 x 45 cm. [450]

Physical geog., geology, climate, mineral resources, forests, population (origins and density), transportation, communications, boundaries, aborigines, explorers' routes (pp. 1–68); city plans (pp. 69–80); diagrams (pp. 81–124).

Burpee, L. J., ed. *An Historical Atlas of Canada: With Introduction, Notes, and Chronological Tables.* Maps by John Bartholomew and Son. Toronto, Nelson, 1927. 32 pl.; 55 pp. 22 x 28 cm.   [451]
Physical conditions and native races, discoveries and explorations, military events, political development, industrial development, boundary disputes and settlements, population. In text, explanatory note and bibl. refs. for each map.

UNITED STATES

(See also above, pp. 35–36, note 37, government documents; [88], guidebooks; [184, 185], early cartography; and [277], regional and city planning.)

Bibliographies: General and Topical

There follows a key to the dates and serial numbers of publications listed in the present volume which contain bibliographical data classified by states (and, in a few instances, by other regional subdivisions) of the United States: climate (Western states, 1935), 475; economic materials in state publications to 1904, 469; geography (1931), 424; geology (1785–1941), 426; glacial geology (1945), 425; German books (1926), 414; historical (1930) (p. 24, note 22, Winchell); (1932), 337 (Vol. 3, 1932, pp. 691–703); (1935), 137; (1916–1938), 411; (1928), 468; historical and geographical society publications (1907), 429; historical geography (1933), 465; land classification (1940), 293; land settlement (1934), 295; land utilization (1938), 294; maps (1938), 97; miscellaneous (1942), 458; (1942), 457; mountain ascents (1945), 416; natural history (1769–1865), 467; Pacific Northwest (1942), 431, (1939), 432; periodical articles (trans-Mississippi West) (1942), 474; "travel reading" (for tourists) (1945), 421; state geological surveys (1932), 466; state publications (1930) (p. 36, note 37, Wilcox); state and regional planning (1935–), 454; geography (1943), 1160.   [452]

Mudge-Winchell [1]: atlases, historical, a 338, d 62; bibliographies, national and trade, a 380–384, b 54, c 75–76, d 71–73; bibliographies, historical, a 361–362, b 52, c 72, d 67–68; dissertations and research projects, a 29–30, b 12, c 14–15, d 12–13; encyclopedias, a 41–43, b 13, d 14–15; gazetteers and geographical dictionaries, a 326; geographical names, a 331–332, b 47, d 60–61; government agencies, a 116–119, b 21, d 23; government documents, a 364–370, b 52, c 73, d 69–70; libraries, catalogues, a 423, d 79; libraries, directories, a 420, c 81, d 76–77; libraries, special collections, a 421–422, c 81; periodicals, bibliographies and directories of, a 18–19, c 10–11; periodicals, indexes, a 6–8, b 9, c 9–10, d 7; societies and foundations, a 35, c 13, d 11–12; statistical yearbooks, abstracts, etc., a 123, c 26, d 25; yearbooks, general, a 121.—American regionalism: Odum and Moore [267], 1938, pp. 643–675.—Lewin [337], Vol. 3, 1932, pp. 691–703.—Bibliography of bibliographies on the Negro in the United States: Work [415], 1928, pp. 630–636.   [453]

Many of the reports of state and regional planning organizations in the United States constitute documents of geographical value. Bibliographies of these will be found in (1) National Resources Board, *State Planning: A Review of Activities and Progress,* Washington, 1935, pp. 294–305; (2) National Resources Committee, *State Planning: Programs and Accomplishments, December 1936,* Washington, 1937, pp. 99–128 (includes references in (1)); and (3) mimeographed supplements to (2) entitled *Bibliography of Reports by State and Regional Planning Organizations,* Nos. 1–[4], 1937– [1940]. See also notes in *Geog. Rev.* [476], 27 (1937), 489–491.   [454]

Fairchild, Wilma B. "The American Scene," *Geog. Rev.* [476], 36 (1946), 291–302.   [455]
Review article on recent "books that describe, portray, or interpret the American countryside."

Coan, O. W., and R. G. Lillard. *America in Fiction: An Annotated List of Novels That Interpret Aspects of Life in the United States.* Rev. ed. Stanford University, Stanford University Press, 1945. 170 pp.   [456]
By topics, a few subcl. areally, ann.; auth. index.

Beers, H. P. *Bibliographies in American History: Guide to Materials for Research.* New York, H. W. Wilson, 1942. 503 pp.   [457]
Deals with works of all kinds that might be of value to the student of history, including many of a geog. nature. The bibls. and aids listed in the chapters on genl. aids, on the United States as a whole, economic

hist., and on the several states, territories, etc., should be of particular value to geographers. By topics (incl. maps) and areas; genl. index.

Logasa, Hannah. *Regional United States: A Subject List.* Boston, Faxon, 1942. 86 pp. (*Useful Reference Series,* No. 69.)   [458]
Covers "ballads, fiction, travel, geographic accounts, biography, autobiography, memoirs, poetry, essays, short stories, tall stories, folklore, history." Alph. by topics, many subcl. by places; index to publishers; areal index.

Platt, Elizabeth T. "Portrait of America: Guidebooks and Related Works," *Geog. Rev.* [476], 29 (1939), 659–664.   [459]
Review article on recent works.

Gaines, S. H., and others. *Bibliography on Soil Erosion and Soil and Water Conservation.* Washington, 1938. 655 pp. (U.S. Dept. of Agric., *Miscellaneous Publ.,* No. 312.)   [460]
Deals predominantly with the United States. Alph. by topics; 4,388 entries; ann.; auth. index.

Culver, Dorothy C. *Land Utilization: A Bibliography.* (Bureau of Public Administration, University of California.) Berkeley, 1935. 231 mimeo. pp. Supplement, 1937, 145 mimeo pp.   [461]
Deals with materials in English on the United States. Cl. by topics under main hdgs.: public lands, land classification, land utilization, land finance, and land protection; 2,887 entries (1935), 1,437 (1937); a few anns.; analytical table of contents with cross refs. but no index.

Wieland, Lillian H. *Bibliography on Soil Conservation.* (United States Department of Agriculture, Soil Conservation Service.) Washington, 1936. 182 mimeo. pp.   [462]
Deals predominantly with the United States, with scattered refs. to other countries. By auths.; topical index.

Bemis, S. F., and Grace G. Griffin. *Guide to the Diplomatic History of the United States, 1775–1921.* (The Library of Congress.) Washington, 1935. 996 pp.   [463]
Of broader value to the geographer than its title might seem to imply. Part I (pp. 1–779) consists of a comprehensive bibl. guide to bibl. aids, special works, printed and manuscript sources, maps, etc., pertaining to the diplomatic hist. and foreign relations of the United States (chron. by periods and topics, many subcl. areally; ann.; 5,318 entries), incl. a chapter on genl. aids to the study of hist., particularly that of the United States (incl. bibls. of maps). Part II (pp. 783–942),

"Remarks on the Sources," presents much bibl. information on printed and manuscript documents of the United States and foreign governments.

Joerg, W. L. G. "Geography and National Land Planning," *Geog. Rev.* [476], 25 (1935), 177–208.   [464]
Summary and critique of reports of National Resources Board and Science Advisory Board, with bibl. refs. and an appendix on agencies concerned with land use.

Semple, Ellen C. *American History and its Geographic Conditions.* Boston, Houghton Mifflin, 1933. Rev. ed.   [465]
Reading lists on pp. 443–505 (by chron. pers. and areas; ann.).

Summary Information on the State Geological Surveys and the United States Geological Survey. Washington, 1932. 136 pp. (*Bull. of the National Research Council,* No. 88.)   [466]
Though not a bibl., gives generalized statements concerning the publs. of the several surveys.

Meisel, Max. *A Bibliography of American Natural History: The Pioneer Century, 1769–1865, the Role Played by the Scientific Societies; Scientific Journals; Natural History Museums and Botanic Gardens; State Geological and Natural History Surveys; Federal Exploring Expeditions in the Rise and Progress of American Botany, Geology, Mineralogy, Paleontology, and Zoology.* Brooklyn, Premier Publishing Co., 1924–1929. 3 vols.   [467]
Vol. 1 contains "an annotated bibliography [alph. by auths.] of the publications relating to the history, biography, and bibliography of American natural history and its institutions, during colonial times and the pioneer century" published up to 1924, with indexes to subjs. and geog. names. Vols. 2 and 3 contain (1) data on the hist. of the institutions which have contributed to the rise and progress of American natural history and were founded between 1769 and 1865, with bibl. analyses of their publs.; (2) a bibl. (chron.) of books, articles, and miscellaneous publs., not listed in (1); (3) indexes of auths. and naturalists and of institutions. The geographer in search of early published material on the United States will find this comprehensive bibl. useful, especially the place index in Vol. 1.

The United States: A Catalogue of Books Relating to the History of Its Various States, Counties, and Cities and Territories. Cleveland, Arthur H. Clark, 1928. 411 pp.   [468]

Unusually comprehensive and scholarly booksellers' catalogue. By states (alph.); ann.; 10,639 entries.

Hasse, Adelaide R. *Index of Economic Material in Documents of the States of the United States.* Washington, 1907–1922. 13 vols. (*Carnegie Instn. of Washington Publ.,* No. 85.) [469]
A separate vol. for each of the following states covering the periods ending with 1904 and beginning with the dates indicated: California (1849), Delaware (1789), Illinois (1809), Kentucky (1792), Maine (1820), Massachusetts (1789), New Hampshire (1789), New Jersey (1789), New York (1789), Ohio (1789), Pennsylvania (1790), Rhode Island (1789), Vermont (1789).

Regional Bibliographies

(See also [431, 432], Pacific Northwest.)

Haferkorn, H. E. *The Mississippi River and Valley: Bibliography Mostly Non-Technical.* Fort Humphreys, Virginia, The Engineer School, 1931. 125 pp. [470]
A second title page reads: "The Mississippi River and Valley: A Bibliography of Books, Pamphlets and Articles in Periodicals, Together with an Index. Prepared under the direction of the Chief of Engineers, U.S. Army, by Mr. H. E. Haferkorn, Librarian, Engineer School Library, Army War College, D.C., July 1930." Sections on books and pamphlets (chron. by topics.) and on period. arts. (chron. by titles); 622 entries; index of auths. and titles.

Ivey, J. E., Jr. *Channeling Research into Education: Prepared for the Committee on Southern Regional Studies and Education.* Washington, 1944. 205 pp. (*Amer. Council on Education Studies,* Ser. 1, *Reports of Committees and Conferences,* No. 19.) [471]
Bibl. of "Selected Source Materials on Southern Resources and Problems" on pp. 129–187: by topics, incl. natural resources, human resources, social resources.

Dobie, J. F. *Guide to Life and Literature of the Southwest, with a Few Observations.* Austin, The University of Texas Press, 1943. 111 pp. [472]
Deals primarily with historical and literary works; 33 short chapters on varied themes (e.g. General Helps, Texas Rangers, Women Pioneers, Mining and Oil, Nature, Wild Life, Naturalists, Pony Express), each followed by bibl. refs.

Tucker, Mary. *Books of the Southwest: A General Bibliography.* New York, Augustin, 1937. 105 pp.   [473]
By topics, incl. exploration, the natural sciences, travel, and description.

Winther, O. O. *The Trans-Mississippi West: A Guide to Its Periodical Literature (1811–1938).* Bloomington, 1942. 278 pp. (*Indiana University Publications: Social Science Series,* No. 3.)   [474]
The material has been selected from periodicals "of a professional or semiprofessional character" but does not include "the special publications by professional societies, such as proceedings and reports of historical societies." Topical (incl. bibls.) and place cl. (alph., chron.); 3,501 entries; auth. index.

U.S. Forest Service Library. *Precipitation and Climate, Particularly as They Affect the Western States: A Bibliography.* [Washington], Nov. 1, 1935. 16 mimeo. pp.   [475]
Topical cl., subcl. by places.

The Principal Geographical Periodicals

(See also [44–46], series.)

The Geographical Review

The American Geographical Society of New York. 1916–. Quarterly. Prior to 1916 the organ of the American Geographical Society was called the *Bulletin of the American Geographical Society* and certain earlier issues were entitled *Transactions, Proceedings,* and *Journal.* These date from 1852. See "Table Showing Composition and Arrangement of the Bulletin of the A.G.S., 1852–1915," p. xi in first Index cited below. The *Bulletin* in its latter years and the *Review* until 1921 were published monthly. Indexes: to the *Bulletin,* 1852–1915; to the *Geog. Rev.,* 1916–1925; 1926–1935; 1936–1945 (forthcoming); see also [530].   [476]
Original arts., critical book reviews, and notes on current publs. and events of geog. interest. Maps and "minor publications" were listed by regions and by topics until January, 1920.

The National Geographic Magazine

Washington, The National Geographic Society. 1889–. Monthly. Index: 1899–1936.   [477]
Emphasis has been laid upon developing the *Magazine* as an instrument for the popularization of geog. During its early years the *Magazine* contained book reviews.

The Journal of Geography: A Magazine for Schools
    Chicago, published by A. J. Nystrom and Co. for the National
Council of Geography Teachers, 1897–. Monthly except June, July,
and August. Entitled *Journal of School Geography,* 1897–1901 inclusive. Index: 1897–1921.   [478]
Devoted to geog. education. Arts. and book reviews.

Annals of the Association of American Geographers
    1911–. Quarterly (annual, 1911–1922). Editor (1947): R. H.
Brown, University of Minnesota, Minneapolis, Minn. Index: 1911–
1935   [479]
Technical arts. by specialists. Abstracts of papers read at meetings of
the Association. No book reviews or bibls.

Economic Geography
    Worcester, Mass., Clark University. 1925–. Quarterly.   [480]
Arts. and book reviews.

Other Geographical Periodicals

Bulletin of the Geographical Society of Philadelphia
    1893–1939. Quarterly. Index: 1893–1939.   [481]
A few book reviews and a brief section devoted to "Geographic News
and Notes."

Bulletin of the Geographic Society of Chicago
    1899–. Irregularly.   [482]

Association of Pacific Coast Geographers: Yearbook
    Cheney, Washington. 1935–. Annual.   [483]

Texas Geographic Magazine
    Dallas, Texas Geographic Society. 1937–. Semiannual.   [484]

*The Professional Geographer: Bulletin of the American Society for
Professional Geographers*
    1943–. Irregularly. Title: *Bull. Amer. Soc. for Geog. Research,*
1943–1944; *Bull. Amer. Soc. for Professional Geographers,* 1944–
1945.   [485]

Maps and Atlases: Bibliographical Data

Official map making activities in the United States are described in
    detail in Thiele [97], 1938, pp. 107–220, 312–338, with many refer-
ences to bibliographical aids to the finding of maps issued by fed-

eral and state agencies, including local, state, and regional planning groups.   [486]

Karpinski, L. C. *Bibliography of the Printed Maps of Michigan, 1804–1880, with a Series of Over One Hundred Reproductions of Maps; Constituting an Historical Atlas of the Great Lakes and Michigan.* Including discussions of Michigan Maps and Mapmakers by W. L. Jenks. Lansing, Mich., Michigan Historical Commission, 1931. 36 pl., 539 pp.   [487]
The reproductions (half tones and line cuts) were also published separately as *Historical Atlas of the Great Lakes and Michigan to Accompany the Bibliography of the Printed Maps of Michigan,* Lansing, Michigan Hist. Comm., 1931, 103 pp. See review in *Geog. Rev.* [476], 23 (1933), 165.

Atlases

(See also [114, 115], general atlases.)

Lord, C. L. and Elizabeth H. *Historical Atlas of the United States.* New York, Holt, 1944. 206 pp. maps (all uncolored, except for violet-buff tint for hill shading on a few). 262 pp. text (pref., contents, statistical tables, index). 21 x 28 cm.   [488]
312 maps in 4 main sections: (1) genl. (physical geog., mineral resources, types of farming, etc., 19 maps); (2) colonial period (26 maps); (3) 1775–1865 (119 maps); (4) 1865–1941 (148 maps). Many topics pertaining to political, military, social, and cultural hist. are represented in this comprehensive but handy and inexpensive atlas.

Adams, J. T., Ed. in Chief, and R. V. Coleman, Managing Ed. *Atlas of American History.* New York, Scribner, 1943. 147 uncolored pl.; 75 pp. (foreword, index, etc.). 18 x 26 cm.   [489]
"In the main, the questions answered by the maps relate to the location of towns, forts, Indian villages, missions, roads, and other cultural features in past times" (from review by R. H. Brown in *Geog. Rev.* [476], 33 (1943), 683–685).

Paullin, C. O. *Atlas of the Historical Geography of the United States.* Ed. by J. K. Wright. Published jointly by Carnegie Institution of Washington and the American Geographical Society of New York, 1932. 166 pl., 162 pp. 27 x 37 cm. (*Carnegie Institution of Washington Publ.,* No. 401.)   [490]
The natural environment (pl. 1–7); reproductions of early maps, 1492–1867 (pl. 8–32); Indians, 1567–1930 (pl. 33–37); explorers' routes,

1535–1852 (pl. 38–39); lands, 1603–1930 (pl. 40–59); population, etc., 1650–1930 (pl. 60–80); colleges, universities, and churches, 1775–1890 (pl. 80–88); boundaries, 1607–1927 (pl. 89–101); political parties and opinion, 1788–1930 (pl. 102–122); political and social reforms, 1775–1931 (pl. 123–132); industries and transportation, 1620–1931 (pl. 133–147); foreign commerce, 1701–1929 (pl. 148–151); distribution of wealth, 1799–1928 (pl. 152–155); city plans, 1775–1803 (pl. 156–159); military history, 1689–1919 (pl. 160–165), world map of various external relationships of the United States (pl. 166). Comprehensive text dealing with sources for and methods used in compilation of the maps. Genl. index. See descriptive article on this atlas in *Geog. Rev.* [476], 22 (1932), 353–360.

Atlas of American Agriculture: Physical Basis, Including Land Relief, Climate, Soils, and Natural Vegetation of the United States. Prepared under the supervision of O. E. Baker. (U.S. Department of Agriculture.) Washington, Govt. Print. Off., 1936. 228 pp. 35 x 48 cm. [491]
The original plan for this Atlas included not only the sections printed in this volume but also other sections relating to a wide variety of other topics. Of these latter only 2 were published: "Cotton," as advanced sheets in 1918, and "Rural Population" in 1919. In the present volume (which lacks a table of contents), the following sections, each paged separately, will be found in the order listed: "Land Relief," by F. J. Marschner, 1936, 6 pp.; "Climate: Temperature, Sunshine, and Wind," by J. B. Kincer, 1928, 34 pp.; "Climate: Frost and the Growing Season," by W. G. Reed, 1918, 13 pp.; "Climate: Precipitation and Humidity," by J. B. Kincer, 1922, 48 pp.; "Soils," by C. F. Marbut, 1935, 98 pp.; "Natural Vegetation" ("Grassland and Desert Shrub," by H. L. Shantz; "Forests," by Raphael Zon), 1924, 29 pp. Each part contains one or more double-page map plates of the United States, 1:8,000,000, in color, and there are many maps and diagrams in the text. The section on soils contains Marbut's magnificent "Soil Map of the United States," 1:2,500,000, in 12 full-page plates.

Statistical Atlas of the United States. Prepared under the supervision of C. S. Sloane. . . . (Department of Commerce, Bureau of the Census.) Washington, Govt. Print. Off., 1925. 476 pp.: 434 pp. uncolored plates, 3 inserted plates in color; 42 pp. text (introduction, contents, index, etc.). 23 x 30 cm. [492]
Population densities and trends (pl. 1–238), agriculture, manufactures, mines and quarries, financial statistics of cities and states, vital statistics, etc. (maps and graphs).

Earlier statistical atlases of the United States were published by the Census Bureau in 1874, 1898, 1903, and 1914.

Hewes, F. W., and Henry Gannett. *Scribner's Statistical Atlas of the United States: Showing by Graphic Methods Their Present Condition and Their Political, Social and Industrial Development.* New York, Scribner, 1883. 151 pl.; 120 pp. 35 x 40 cm.   [493]
Physical geog., political hist., "progress," population, mortality, education, religion, occupations, finance and commerce, agriculture, livestock, manufactures, etc. (maps and graphs). Many colored statistical maps with data plotted by counties.

MEXICO

(See also [91], guidebook.)

Mudge-Winchell [1]: bibliographies, a 406–407, c 79–80, d 74; bibliographies, historical, a 358, b 51, c 71, d 66; commercial handbook (1923), a 150; geographical and historical dictionaries, a 358, d 66; yearbooks, a 127, d 25.—The series *Monografías bibliográficas mexicanas,* published by the Secretaria de Relaciones Exteriores under the direction of Genaro Estrada, contains bibliographies of the states of Coahuila (No. 10, 1927), Michoacán (No. 25, 1932), Morelos (No. 27, 1933), Sinaloa (No. 6, 1926), Tabasco (No. 16, 1930), and Zacatecas (No. 26, 1932). The continuation of this series, *Bibliografías mexicanas,* published by Departamento Autónomo de Prensa y Publicidad, contains bibliographies of Baja California (No. 2, 1937), Quintana Roo (No. 3, 1937), and Vera Cruz (No. 1, 1937). For references to these and other regional bibliographies see Millares, etc. [496], pp. 96–100.   [494]

Beals, Ralph, Robert Redfield, and Sol Tax. "Anthropological Research Problems with Reference to the Contemporary Peoples of Mexico and Guatemala," *American Anthropologist,* Vol. 45, No. 1 (1943).   [495]
Bibl. on pp. 17–21 (place cl., subcl. by auths.).

Millares Carlo, Agustín, and José Ignacio Mantecón. *Ensayo de una bibliografía de bibliografías mexicanas.* (Departamento del Distrito Federal, Dirección de Acción Social, Oficina de Bibliotecas.) Mexico, 1943. 240 pp.   [496]
Part I, general bibls. of America that deal with Mexico; Part II, bibls.

of Mexico, with a section on regional bibls. (by states). By forms of
publ. and by topics; genl. index.

Ker, Annita M. *A Survey of Mexican Scientific Periodicals, to Which
Are Appended Some Notes on Mexican Historical Periodicals.*
Baltimore, 1931. 120 pp. (*Publications of the Harvey Bassler Foun-
dation.*) [497]
Part I, "An Account of the Societies and Institutions which are Pub-
lishing Scientific Periodicals in Mexico at Present," pp. 1–37; Part II,
"Bibliography," pp. 38–96 (alph. by titles; ann.; topical index).

Nelson, E. W. *Lower California and Its Natural Resources.* Constituting *Memoirs of the* [*U.S.*] *National Acad. of Sciences,* Washington, Vol. 16, Memoir 1, 1921. [498]
Bibl. on pp. 147–171 (chron. by titles; ann.).

Periodicals

Boletín de la Sociedad Mexicana de Geografía y Estadística
Mexico City. 1839–. Irregularly. Index: 1839–1918. [499]

Magazine de Geografía Nacional
Mexico. 1925–1926. Monthly. [500]

Boletín de la Sociedad Michoacana de Geografía y Estadística
Morelia. 1905–1912. Semimonthly; monthly. [501]

CENTRAL AMERICA

Mudge-Winchell [1]: bibliographies, Costa Rica, a 395; bibliographies, Guatemala, c 79; bibliographies, Honduras, c 79, d 74; geographical dictionary, Salvador, d 61; government documents, a 374;
statistical yearbook, Nicaragua, c 26. [502]

Phillips, P. L. *A List of Books, Magazine Articles, and Maps Relating
to Central America, Including the Republics of Costa Rica, Guatemala, Honduras, Nicaragua, and Salvador, 1800–1900.* (Bureau of
the American Republics.) Washington, 1902. 109 pp. [503]
Books and period. arts. (alph. by auths.) and maps (chron. by titles)
cl. under hdgs. for Central America as a whole and for each republic.

Dobles Segreda, Luis. *Indice bibliográfico de Costa Rica.* San José,
Costa Rica, Imprenta Lehmann, 1927–1936. 9 vols. published, 3
others contemplated. [504]

By topics, incl. agriculture, hist., geog. and geology. The section "Geografía y Geología" (Vol. 2, pp. 134–608) includes detailed analyses of books and period. arts., 1836–1927 (chron., 365 entries, ann.), and a list of maps, 1523–1926 (chron., 259 entries, ann.).

Brown, Ann D. *The Panama Canal and the Panama Canal Zone: A Selected List of References.* (The Library of Congress, Division of Bibliography.) Washington, 1943. 57 mimeo. pp.   [505]
Topical cl. (system.); 430 entries; auth. index with a few topical captions.

_____

Periodicals

Anales de la Sociedad de Geografía e Historia de Guatemala
1924–. Quarterly.   [506]

Maps: Bibliographical Data

Maps in the Library of the Middle American Research Institute. New Orleans, Middle American Research Institute, Tulane Univ., 1941. 285 mimeo. pp. (*An Inventory of the Collections of the Middle American Research Institute,* No. 4.)   [507]
Regional cl.; 889 entries; index of cartographers, printers, and publishers; topical index, subcl. by geog. names. See also *Maps in the Frederick L. Hoffman Collection,* New Orleans, 1939 (No. 3 of same series).

The Caribbean Region

Bibliography of bibliographies on the West Indies: Work [415], 1928, pp. 655–658.—Martineau [331], 1932, pp. 375–390.   [508]

Catalogue of Books, Manuscripts, etc. in the Caribbeana Section (Specializing in Jamaicana) of the Nicholas M. Williams Memorial Ethnological Collection. (Boston College Library.) Chestnut Hill, Mass., 1932. 133 pp.   [509]
By topics (discovery, piracy, slavery, genl. works) and places (Jamaica, Haiti, and other regions of Caribbean area); 1,674 entries.

Trelles, C. M. "Bibliografía antillana," *Hispanic Amer. Hist. Rev.,* 4 (1921), 324–330.   [510]
Brief comment and list of West Indies bibls.

List of Works Relating to the West Indies. (New York Public Library.) (Reprinted from *Bull. of the New York Public Library,* Vol. 16 (1912), Nos. 1–8.) New York, 1912. 392 pp. [511]
A section (37 pp.) devoted to bibls., genl. history, description, etc. (alph. by auths.) and sections on works relating to individual islands and colonies (mostly subcl. under hdgs. for periods. and genl. works; the longer sections also subcl. in considerable detail by subjs. and, in the case of Cuba, subcl. by place); auth. and subj. index.

Cundall, Frank. *Bibliography of the West Indies (Excluding Jamaica).* Kingston, The Institute of Jamaica, 1909. 177 pp. [512]
In addition to the West Indies covers Central America, Panama, Florida, Colombia, Venezuela, the Guianas, and also British West Africa. Books and pamphlets (but not period. arts.) cl. areally, with titles chron.; 2,889 entries; list of parliamentary papers (chron.); index of names (auths., cartographers, and subjects of memoirs).

———

Platt, R. R., and others. *The European Possessions in the Caribbean Area.* New York, 1941. (*Amer. Geog. Soc. Map of Hispanic America,* Publ. No. 5.)
List of refs. on pp. 101–106. [513]

Brown, Ann D. *British Possessions in the Caribbean Area.* (The Library of Congress, Division of Bibliography.) Washington, 1943. 192 mimeo. pp. [514]
Covers bibls., description and travel, geology and natural resources, history, economic and social conditions, politics, govt., and international relations. Topical and place cl. (system., alph.); 1,487 entries; auth. and subj. indexes.

Hiss, P. H. *A Selective Guide to the English Literature on the Netherlands West Indies: With a Supplement on British Guiana.* New York, The Netherlands Information Bureau, 1943, 142 pp. (*Booklets of the Netherlands Information Bureau,* No. 9.) [515]
Topical cl. under 3 main hdgs. "Curaçao," "Surinam," and "British Guiana"; auth. index.

Benjamins, H. D., and J. F. Snelleman. *Encyclopaedie van Nederlandsch West-Indië.* The Hague (Nijhoff) and Leiden (Brill), 1914–1917. 792 pp. [516]
This extensive regional gazetteer includes bibl. notes appended to many of the arts. (see especially the art. "Aardkunde").

CUBA

(See also [91], guidebook.)

Mudge-Winchell [1]: bibliographies, a 395, c 79, d 73; encyclopedia, d 64.   [517]

Trelles, C. M. "Biblioteca geográfica cubana (de 1914 a 1939)," *Revista de la Sociedad Geográfica de Cuba,* 13, número extraordinario (1940), 107–150.   [518]
Continuation (with same system of cl.) of second section of [519].

Trelles, C. M. *Biblioteca geográfica cubana.* Matanzas, J. F. Oliver, 1920. 344 pp.   [519]
2 main sections: (1) geog. works which, while not dealing with Cuba, are of special interest to Cubans (pp. 1–59) and (2) works on the geog. of Cuba (pp. 60–308). Each section cl. by topics (system.), incl. history of geog., mathematical, physical, human, biological, and military geog. cartography, guidebooks, etc.; ann.; auth. and subj. indexes. (Cl. based on that of *Bibliotheca geographica* [10].

Quesada, Gonzalo de. *Cuba.* (International Bureau of the American Republics.) Washington, 1905.   [520]
Chapter 17, pp. 315–512, of this handbook is a bibl. comprising lists, compiled by A. P. C. Griffin, of books relating to Cuba (by auths., alph.), learned society publs. (alph.), period. arts. (chron.), and govt. docs. (chron.), and a list of maps (chron.) compiled by P. L. Phillips.

Periodical

Revista de la Sociedad Geográfica de Cuba
          Havana. 1927–. Quarterly (irregularly, 1927–1940).   [521]

HAITI

Mudge-Winchell [1]: bibliography, d 74.   [522]

Cabon, Adolphe. "Notes bibliographiques sur la géographie d'Haïti," *Bull. semestriel de l'observatoire météorologique du Séminaire-Collège St. Martial, Port-au-Prince, Haïti,* July–Dec., 1916, pp. 149–174; *Bull. annuel* of the same, 1917, pp. 93–128.   [523]
Descriptive comment on the geog. literature of Haiti from its discovery by Columbus until the time of publication. The first part, covering the period before 1789, is arranged chron.; the second, which deals with the subsequent period, is cl. by topics, incl. physical, economic, and human geog., population, statistics, etc.

Roth, H. L. "Bibliography and Cartography of Hispaniola," *Supplementary Papers of the Royal Geog. Soc.,* Vol. 2, pp. 41–97. London, 1889.   [523a]
Sections on: (1) books and a few period. arts., by auths. (alph.); (2) anonymous pubs.; (3) maps (chron.); ann.; chron. and subj. indexes.

Periodical

Revue de la Société d'Histoire et de Géographie d'Haïti
   Port-au-Prince. 1925–. Quarterly (formerly irregularly). Prior to 1930 (?) entitled *Bulletin.*   [524]

JAMAICA

Cundall, Frank. *Bibliographia Jamaicensis: A List of Jamaica Books and Pamphlets, Magazine Articles, Newspapers, and Maps, Most of Which Are in the Library of the Institute of Jamaica.* Kingston, Jamaica, The Institute of Jamaica, 1902. 83 pp. *Idem, Supplement to Bibliographia Jamaicensis.* Kingston, 1908. 38 pp.   [525]
The main bibl. and the *Supplement* each contain sections on: (1) books and pamphlets cl. by topics (incl. hist., description, guidebooks, natural history, works of reference) and (2) period. arts. similarly cl. The main bibl. also includes a section on maps and town plans (chron.).

PUERTO RICO

Brown, Ann D. *Puerto Rico: A Selected List of Recent References.* (The Library of Congress, Division of Bibliography.) Washington, 1943. 44 mimeo. pp.   [526]
By topics, incl. bibls., genl. works, description and travel, economic and social conditions, industry, resources, commerce; 381 entries; auth. index.

Pedreira, A. S. *Bibliografía puertorriqueña (1493–1930).* Madrid, Hernando, 1932. 739 pp. *(Monografías de la Universidad de Puerto Rico,* Series A: *Estudios hispánicos,* No. 1.)   [527]
By topics, incl. bibls., genl. information, natural history (incl. physical geog.), health, economic and social conditions, hist., culture, literature; auth. index.

MARTINIQUE, VIRGIN ISLANDS

Blanchard, L. R. *Martinique: A Selected List of References.* Supplementing a list prepared by the Division of Bibliography in 1923.

(The Library of Congress, Reference Dept.) Washington, 1942. 57 mimeo. pp.   [528]
"Emphasis has been placed on history, description and travel, economics and politics." By topics, incl. bibls., maps, Mt. Pelée, etc.; ann.; 257 entries; auth. index.

Reid, C. F., ed. *Bibliography of the Virgin Islands of the United States*. New York, H. W. Wilson, 1941. 241 pp.   [529]
One of the "Bibliographies of the Territories and Outlying Possessions of the United States" prepared with the assistance of the Federal Works Agency, Work Projects Administration for the City of New York. Alph. by many topics, incl. bibls., geog., geology, maps; ann.; auth. index.

Latin America

(See also [172], voyages of exploration.)

Mudge-Winchell [1]: bibliographies, a 411–412, c 78–79, d 73; commercial guidebooks, c 30; commercial handbooks, a 150; statistical yearbooks, d 23.—During the Second World War there was established at Yale University the Strategic Index to the Americas, a project of which "the purpose is to assemble and to classify systematically basic data on the other American republics."—Review article by R. R. Platt on 17 recent books on Latin America, in *Geog. Rev.* [476], 33 (1943), 304–311.—Notes on recent bibliographies on Latin America, in *Geog. Rev.* [476], 27 (1937), 334–336.— A. C. Wilgus has published in the *Hispanic American Historical Review* [545] lists of articles bearing on Latin America in (1) periodicals of the American Geographical Society [476], 1852–1933 (*Hisp. Amer. Hist. Rev.*, 14 (1934), 114–130); (2) the *National Geographic Magazine* [477], 1888–1932 (*ibid.*, 12 (1932), 493–502); and (3) the *Geographical Journal* [625], 1893–1934 (*ibid.*, 19 (1939), 117–126).   [530]
A "tentative summary" of facts of value to librarians and students concerning the executive departments of the Latin American countries, rather than a bibliography, is provided in J. B. Childs' "Bibliography of Official Publications and the Administrative System in Latin American Countries," (*Inter-American Bibl. and Libr. Assn. Publs.*, Ser. 2, Vol. 1, New York, H. W. Wilson, 1938, pp. 131–172). See also [532].   [531]

Lesser, Alexander. *Survey of Research on Latin America by United States Scientists and Institutions.* (National Research Council: The Committee on Latin American Anthropology, Division of Anthropology and Psychology.) Washington, 1946. 133 mimeo. pp. [531a]
Not a bibl. but, rather, a survey of current projects, programs, and trends in the fields of anthropology, geog., sociology, social welfare, economics, hist., political science, statistics, and demography.

Childs, J. B., general ed. *A Guide to the Official Publications of the Other American Republics.* (The Library of Congress.) Washington, no date, [1945–?]. I. *Argentina* (124 pp.); II. *Bolivia* (66 pp.); VII. *Cuba* (40 pp.), constituting *Latin American Series* Nos. 9, 10, 11, respectively. [532]
This comprehensive *Guide* is being issued in paper-bound parts numbered in alph. order by countries (the *Latin American Series* of which they form parts being numbered in order of publication). In each part the entries are cl. by the agencies that issue them and there is an index to names and titles with a few subj. hdgs. (e.g. "Maps"). See also [531].

Shelby, Charmion, ed. *Latin American Periodicals Currently Received in the Library of Congress and the Library of the Department of Agriculture.* (The Library of Congress.) Washington, 1944. 256 pp. (*Hispanic Foundation, Latin American Series,* No. 8.) [533]
Alph. by titles of periods.; 1,578 entries; ann.; index by countries; topical index. "Selected Bibliography" of bibls. of periods., pp. 218–222.

Hopper, Elizabeth G. *A Preliminary List of Latin American Periodicals and Serials.* Washington, 1943. 195 mimeo. pp. (U.S. Dept. of Agric., *Library List,* No. 5.) [534]
Alph. by areas, subcl. by institutions. The periodicals listed are for the most part agricultural.

Wilgus, A. C. "Recent Publications in the United States Dealing with Latin American Affairs and Related Fields," in *Inter-American Bibl. Rev.,* [545a], 1–3, (1941–1943). [535]
In 1941 covered books and pamphlets printed in English. Topical cl. (system.); ann.

James, P. E. *Latin America.* New York, Odyssey Press. 1942. [536]
List of refs. on pp. 869–886 to works which have appeared since 1930; by regions; 293 entries.

Jones, C. K. *A Bibliography of Latin American Bibliographies.* 2d ed. (The Library of Congress, Hispanic Foundation.) Washington, 1942. 311 pp.   [537]
By countries (alph.); 3,016 entries; ann.; genl. index.

Handbook of Latin American Studies: A Selective Guide to the Material Published in [year] *on Anthropology, Archives, Art, Economics, Education, Folklore, Geography, Government, History, International Relations, Labor and Social Welfare, Language and Literature, Law, Libraries, Music and Philosophy.* (Joint Committee on Latin American Studies of the National Research Council, the American Council of Learned Societies, and the Social Science Research Council.) Cambridge, Harvard Univ. Press, 1935-. Annual. Title and name of sponsor vary.   [538]
Nos. 1-4 edited by Lewis Hanke (Raul d'Eça, associate editor of No. 4); No. 5, by Lewis Hanke and Miron Burgin; Nos. 6 and 7, by Miron Burgin.
  Comprehensive coöperative work. The sections on geog. and cartography have been contributed by well-known geographers of the United States. Topical cl., subcl. by places; *ca.* 6,000 entries (1941); indexes: by auths. and geog. names.

Humphreys, R. A. *Latin America.* London, Royal Institute of International Affairs, 1941. 36 pp. (*Chatham House Bibliographies.*)   [539]
A "selected reading list" of books in English cl. by topics (incl. hist., economics, international relations) and countries; ann.

U.S. Coördinator of Inter-American Affairs. *Preliminary Listing of Publications Dealing with Latin America Issued by the Agencies of the United States Government During the Past Five Years.* Washington, Sept. 15, 1941. 25 mimeo. pp.   [540]
By govt. agencies (alph.), subcl. by topics.

Vivó, J. A. *Catalogo de la Biblioteca (1930-1939).* Tacubaya, Mexico, 1940. 412 pp. (*Instituto Panamericano de Geografía e Historia,* Publ. No. 47.)   [541]
Catalogue of the Library of the Pan-American Institute of Geography and History, a general library relatively strong in geog., cartographic, and historical material on Latin America. By places, subcl. by topics, incl. maps; auth. index; index of maps.

Harvard University: Bureau for Economic Research in Latin America. *The Economic Literature of Latin America: A Tentative Bibliography.* Cambridge, Mass., Vol. 1, 1935, 332 pp.; Vol. 2, 1936, 366 pp.   [542]
Sections on Latin America as a whole in both vols. The remainder, comprising the greater parts of Vols. 1 and 2, deals respectively with South America and with Mexico, Central America, and the West Indies. Each vol. contains a survey of statistical sources. Vol. 2 contains notes on collections of Latin American economic literature in leading libraries. Alph. by countries, subcl. by topics; 12,520 entries; auth. index.

Theses on Pan American Topics Prepared by Candidates for Degrees in Universities and Colleges in the United States. Washington, 1933. 113 mimeo. pp. (*Pan American Union Bibliographic Series,* No. 5, 2d ed.)   [543]
Auth. cl.; 1,111 entries; subj. index.

Quelle, O. "Das romanische Amerika, 1913–25," *Geog. Jhrb.* [6], 41 (1926), 360–424.   [544]
Areal cl., with a section on Latin America as a whole; 1,114 entries; ann.; auth. index.

Periodicals

The Hispanic American Historical Review
Baltimore. 1918–. Quarterly.   [545]
Book reviews and a list of recent publs. on Hispanic American hist., geog., and political and economic conditions.

Inter-American Bibliographical Review
Washington, Inter-American Bibliographical and Library Association. 1941–1943. Quarterly.   [545a]
Bibls. and arts. on bibl. subjs.

Ibero-Amerikanisches Archiv
Berlin (formerly Bonn), Ibero-Amerikanisches Institut. 1924– [1939]. Quarterly. Subtitle 1924–1930: *Zeitschrift des Ibero-Amerikanischen Forschungsinstitut der Universität Bonn.* Succeeds: *Mitteilungen des Deutsch-Südamerikanischen und Iberischen Instituts,* Stuttgart and Berlin, 1913–1921, irregularly (title varies).   [546]

Maps: Bibliographical Data

Jones [537], 1942, index, under "Maps."—The *Handbook of Latin-American Studies* [538], 1935–1941, contains a section devoted to cartography.—Official cartography: Thiele [97], 1938, pp. 232–241.—On the American Geographical Society's *Map of Hispanic America* on the scale of 1:1,000,000: *Geog. Rev.* [476], 36 (1946), 1–28.
[547]

Wilgus, A. C. *Maps Relating to Latin America in Books and Periodicals.* Washington, 1933. 119 mimeo. pp. (*Pan American Union Bibliographic Series,* No. 10.)
[548]
4 main parts: (1) Backgrounds (with subhdgs. covering geog., ethnography, and European background), (2) The Colonial Period, (3) The Period of Revolutions (1808–1824), (4) The Modern Period. See *Geog. Rev.* [476], 27 (1937), 335.

A Catalogue of Maps of Hispanic America, Including Maps in Scientific Periodicals and Books and Sheet and Atlas Maps, with Articles on the Cartography of the Several Countries and Maps Showing the Extent and Character of Existing Surveys. New York, 1930–1933. 4 vols. (*Amer. Geog. Soc. Map of Hispanic America Publ.,* No. 3.)
[549]
For maps in books, sheet and atlas maps, and historical maps indications are given as to whether they may be found in the following libraries: Library of Congress, New York Public Library, Pan American Union, American Geographical Society, Columbia, Harvard, and Yale universities. Titles (chron.) cl. by countries, subcl. by forms of publ.; topical index, subcl. by countries. See article by R. R. Platt in *Geog. Rev.* [476], 23 (1933), 660–663.

South America

The introduction to Arredondo [597], 1929, pp. 433–447, presents a brief summary of available bibliographical sources for South America.—Geomorphology: Spreitzer [214], 1938, pp. 251–252.—Soils: Giesecke [230], 1939, pp. 282–284.
[550]

Berninger, O. "Südamerika (1927–38)," *Geog. Jhrb.* [6], 54, Pt. 2 (1939), 449–554; 55, Pt. 2 (1940), 633–759; 56, Pt. 1 (1941), 195–252.
[551]
2 main sections, (1) South America and Latin America (*Iberoamerika*)

in genl. (topical cl.); (2) regions, subcl. by topics; 3,370 entries; ann.; auth. index.

Jones, C. F. *South America*. New York, Holt, 1930. [552]
The list of refs. (pp. 725–765) contains many titles of studies by professional geographers. By topics (incl. major physiographic regions, climates, natural vegetation, transportation, people) and by areas.

Welch, Margaret M. *Bibliography on the Climate of South America*, 1921. 42 pp. (U.S. Weather Bureau: *Monthly Weather Review*, Suppl. No. 18.)
By countries; brief anns. [553]

Catalogo da exposição de geographia sul-americana realizada pela Sociedade de Geographia do Rio de Janeiro. . . . Rio de Janeiro, 1891. 493 pp. [554]
Books and maps dealing with all parts of South America, but more especially with Brazil, cl. by topics (incl. physical geog., geology, mineralogy, anthropology, botany and zoology, orography, political geog., hydrography, commercial geog., mathematical geog., meteorology, etc.) subcl. by places; 1,769 entries, ann.

NORTHERN SOUTH AMERICA

COLOMBIA

Mudge-Winchell [1]: bibliographies, a 394–395; commercial handbook (1921), a 150. [555]

Periodicals

Boletín de la Sociedad Geográfica de Colombia (Academia de Ciencias Geográficas)
Bogotá. Series 2, 1934–. Irregularly. [556]

Revista geográfica de Colombia
Bogotá, Instituto Geográfico Militar. 1936–1939. [557]

VENEZUELA

Mudge-Winchell [1]: commercial handbook (1922), a 150; statistical yearbook, c 26, d 25. [558]

Dalton, L. V. *Venezuela*. London, Unwin, 1912. [559]
Bibl. on pp. 287–313; topical and place cl.; 411 entries.

Sanchez, M. S. *Bibliografía venezolanista: contribución al conocimiento de los libros extranjeros relativos á Venezuela y sus grandes hombres publicados ó reimpresos desde el siglo XIX.* Caracas, Empresa El Cojo, 1914. 494 pp.   [560]
Covers works in Spanish, French, German, and English on geog., ethnography, hist., geology, etc.; alph. by auths.; ann.; no index. For "Adiciones" see *Boletín de la Biblioteca Nacional de Venezuela,* No. 43 (1936), pp. 239-254.

THE GUIANAS

Lewin [337], Vol. 3, 1932, pp. 608-631.—See also [508, 512-516].   [561]

WEST COAST COUNTRIES

ECUADOR

Mudge-Winchell [1]: bibliography, historical, a 398.   [562]

Paz y Miño, L. T. "Bibliografía geográfica ecuatoriana," *Boletín de la Biblioteca Nacional de Quito,* N.S. (1927), No. 10, pp. 178-200; No. 11, pp. 234-278. (Also published separately in *Publicaciones de la Biblioteca Nacional de Quito: Estudios Geográficos,* Quito, 1927. 69 pp.)   [563]
2 sections on works relating to Ecuador, each cl. by auths.: (1) publs. by Ecuadorians, 242 entries; (2) publs. by others, 509 entries.

PERU

Mudge-Winchell [1]: bibliography, c 80, d 74; commercial handbook (1925), a 150; dictionaries, geographical, a 326; dictionaries, historical, a 359; statistical yearbook, a 127.   [564]

Broggi, J. A. "Bibliografía geológica, botánica y zoológica del Perú, 1929-1933," *Boletín de la Sociedad Geológica del Perú,* 6 (1934), 23-123.   [565]
Topical cl.; fully ann.

Periodical

Boletín de la Sociedad Geográfica de Lima
1891-. Quarterly. Index: 1891-1941.   [566]
In the main the arts. are devoted to Peru.

BOLIVIA

Mudge-Winchell [1]: bibliographies, a 393; commercial handbook (1921), a 150; geog. dictionary, a 323. [567]

Periodicals

Boletín de la Sociedad Geográfica de La Paz
1898–1931, 1941–. Irregularly. [568]

Boletín de la Sociedad Geográfica de Sucre
1898–. [569]

CHILE

Mudge-Winchell [1]: bibliographies, a 393–394, c 79; gazetteer, a 324; statistical yearbook, a 125. [570]

Steffen, Hans. "Der Anteil der Deutschen an der geographischen und geologischen Erforschung der Republik Chile während des ersten Jahrhunderts ihres Bestehens," *Deutsche Arbeit in Chile,* constituting *Festschrift des Deutschen Wissenschaftlichen Vereins zu Santiago zur Centenarfeier der Republik Chile (Verhandlungen des Vereins,* Vol. 5, Nos. 3–6), 1 (1910), 188–245. [571]
Running discussion of German contributions to the geog. investigation of Chile with bibl. refs. in the footnotes.

Reiche, Karl. *Grundzüge der Pflanzenverbreitung in Chile.* Leipzig, 1907. Spanish translation by Gualterio Looser: "Geografía botánica de Chile," *Revista chilena de historia y geografía* [575], 62 (1929), 127–205. [572]
In the bibl. (pp. 164–205 of translation) are listed (alph. by auths.) books, period. arts., and maps bearing on the flora of Chile and also certain genl. works on Chile dealing incidentally with plant geog. The translation does not carry the bibl. later than 1907.

Phillips, P. L. *A List of Books, Magazine Articles, and Maps Relating to Chile.* (International Bureau of the American Republics.) Washington, 1903. 110 pp. [573]
Sections on (1) books (auth. cl.); (2) period. arts. (by auths. and a few subjs.); (3) maps (chron.).

Anrique R., Nicolas, and L. Ignacio Silva A. *Ensayo de una bibliografía histórica i jeográfica de Chile.* Santiago de Chile, Imprenta Barcelona, 1902. 698 pp. [574]

2 main parts: hist. (276 pp.) and geog. (400 pp.), each subcl. by topics and each with separate auth. index.; a few anns.

Periodical

Revista chilena de historia y geografía
Santiago de Chile, Sociedad Chilena de Historia y Geografía. 1911-. Now semiannual. Index: 1911–1924.   [575]
Arts. and a few long signed book reviews.

BRAZIL

Mudge-Winchell [1]: atlas, statistical, d 24; bibliographies, a 393, c 79, d 74; bibliographies, historical, a 348; encyclopedia, a 45; gazetteers, a 324; statistical yearbooks, a 124, c 25–26.   [576]

Gorham, Rex. "The Folkways of Brazil: A Bibliography," ed. by Karl Brown, *Bull. of the New York Public Library,* 47 (1943), 255–272, 427–435; 48 (1944), 435–440, 501–511.   [577]
Sections on (1) genl. works on folklore, religion, and anthropology; (2) antecedents in Portugal; (3) Brazilian folklore literature; (4) travel, hist., and tradition; (5) fiction; (6) music and poetry; (7) religious fanaticism; (8) the Negro in Brazil. Entries in each section by auths.; fully ann.; 344 entries.

Barros Paiva, Tancredo de. "Bibliographia do clima brasilico," *Boletim do Ministerio da Agricultura, Industria e Commercio,* Anno 17, Vol. 1 (1928), pp. 638–649, 807–826; also issued separately. [578]
Auth. cl.

Garcia, Rodolpho. "Bibliographia geographica brasileira," *Revista do Instituto Historico e Geographico Brasileiro* [585], Tomo 85, Vol. 139, (1919), pp. 5–105.   [579]
Based on library of the Institute. Titles (chron.) under hdgs. for genl. works, works on Brazil, Brazilian international and interstate frontiers, maps, manuscripts; no refs. to period. arts.

Scherrer, Joseph. "Historisch-geographischer Katalog für Brasilien (1500–1908)," *Annaes da Bibliotheca Nacional do Rio de Janeiro,* 35 (1913), 313–418.   [580]
Almost exclusively German publs., cl. by auths.; topical index with a few place hdgs.

Phillips, P. L. *A List of Books, Magazine Articles, and Maps Relating to Brazil, 1800–1900.* A Supplement to the *Handbook of Brazil*

(1901) compiled by the Bureau of the American Republics. Washington, 1901. 145 pp.   [581]
Sections on (1) books (auth. cl.); (2) period. arts. (by auths. and a few subjs.); (3) maps (chron.).

———

Documentação bibliográfica da Amazônia," *Revista brasileira de geografia* [583], 4 (1942), 433–461, 679–705.   [582]
Sections on books, period. arts., and regional materials (Arquivó corográfico), each by auths.

———

The Principal Geographical Periodicals

Revista brasileira de geografia
 Rio de Janeiro, Instituto Brasileiro de Geografia e Estatística. 1939–. Quarterly.   [583]
Long scholarly arts. and lists of Brazilian geog. publs. Published under the aegis of the Conselho Nacional de Geografia and the Conselho Nacional de Estatística. (See also [584] and *Geog. Rev.* [476], 29 (1930), 506–507.)

Boletim geográfico
 Rio de Janeiro, Instituto Brasileiro de Geografia e Estatística : Conselho Nacional de Geografia. 1943–. Monthly. First 3 numbers entitled *Boletim do Conselho Nacional de Geografia.*   [584]
Revista do Instituto Histórico e Geográfico Brasileiro
 Rio de Janeiro. 1839–. Semiannual. Indexes: 1839–1906, in 68 (1907), 283–435; 1839–1921, publ. separately.   [585]

Other Geographical Periodicals

Revista do Instituto Arqueológico, Histórico, e Geográfico Pernambucano
 1863–. Annual.   [586]
Revista da Sociedade de Geografia do Rio de Janeiro
 1886–. Normally annual. A *Boletim* was published in 1885. The number for 1909–1911 was published in 1919.   [587]
Revista do Instituto Geográfico e Histórico da Bahia
 1894–. Annual.   [588]
Revista do Instituto Histórico e Geográfico de São Paulo
 1895–. Annual.   [589]
Revista do Instituto Histórico e Geográfico Paraibano
 Paraíba. 1909. Irregularly. Not published 1912–1922.   [590]

182 *Regional Aids and Periodicals*

Revista do Instituto Geográfico e Histórico do Amazonas
Manaos. 1917–. [591]
Revista do Instituto Histórico e Geográfico do Rio Grande do Sul
Pôrto Alegre. 1921–. Quarterly. [592]
Boletim da Associação dos Geógrafos Brasileiros (São Paulo)
São Paulo. 1935–. Irregularly. Issued 1935–1936 as *Geografia: Publicação Trimestral da Associação dos Geógrafos Brasileiros.*
(Not published 1937–1942.) [593]
Revista do Instituto Histórico e Geográphico do Pará
1917–1936. Irregularly. Index: 1917–1931, in Vol. 7. [594]

SOUTHEASTERN SOUTH AMERICA

PARAGUAY

Decoud, J. S. *A List of Books, Magazine Articles, and Maps Relating to Paraguay: Books, 1638–1903; Maps, 1599–1903.* A Supplement to the *Handbook of Paraguay,* September, 1902. Washington, International Bureau of the American Republics, 1904. 53 pp. [595]
Sections on (1) books (auth. cl.); (2) period. arts. (auth. cl.); (3) maps (chron.).

URUGUAY

Mudge-Winchell [1]: dictionaries, geographical, a 327; dictionaries, historical, a 363. [596]

Arredondo, Horacio. "Bibliografía uruguaya (contribución)," *Revista del Instituto Histórico y Geográfico del Uruguay,* 6, No. 2 (1929), 433–610. [597]
Lists approximately 450 titles (chron.) issued between 1559–1865, some of which have a geog. bearing.

Periodical

Revista del Instituto Histórico y Geográfico del Uruguay
Montevideo. 1920–[1941–42]. Irregularly. [598]

ARGENTINE REPUBLIC

Mudge-Winchell [1]: bibliographies, a 392, c 79, d 73; bibliographies, historical, c 68; geographical and historical dictionaries, a 323, 348; guide to periodicals, a 20. [599]

Sparn, Enrique. *Bibliografía meteorológica y climatológica de la República Argentina*. Cordoba, 1923. 56 pp. (*Academia Nacional de Ciencias Miscelanea*, No. 7.) Also published in *Revista de la Universidad Nacional de Córdoba*, 10 (1923), 91–140.   [600]
Alph. by auths.; index to geog. names. The Appendix (14 pp.), entitled "Literatura sobre meteorología y climatología de las regiones antártica y sub-antártica americanas" (alph. by auths.) is a list of refs. on the meteorology and climatology of Tierra del Fuego, the Falkland and South Georgia groups of islands, and the Antarctic continent and adjacent islands immediately to the south.

Kühn, Franz. "Deutsche geographische Arbeit in und über Argentinien: Eine systematische Zusammenstellung deutscher Forschung auf dem Gebiete der argentinischen Landeskunde," Part 1, 1850–1900, in *Zeitschrift des Deutschen Wissenschaftlichen Vereins zur Kultur- und Landeskunde Argentiniens*, 1916, pp. 261–275, 286–310; also published separately, Buenos Aires, 1917.   [601]
By topics, incl. expeditions, maps, surveys, geology, hydrography, climate, vegetation, human geog., regional studies; 357 entries; auth. and subj. indexes.

———

Rudolph, W. E. "Southern Patagonia as Portrayed in Recent Literature," *Geog. Rev.* [476], 24 (1934), 251–271.   [602]

Periodicals

Gaea

Buenos Aires, Sociedad Argentina de Estudios Geográficos. 1922–. Irregularly.   [603]
Revista geográfica americana
Buenos Aires, Sociedad Geográfica Americana. 1933–. Monthly.   [604]
Much material relating to the Argentine; also a review and record section.

Anuario geográfico argentino
Buenos Aires, Comité Nacional de Geografía. 1941–.   [605]
Anuario del Instituto Geográfico Militar de la República Argentina
Buenos Aires. 1912–[1929–1932]. Publ. was suspended after 1914 and resumed in 1922 with the number for 1915–1916.   [606]
Boletin del Instituto Geográfico Argentino
Buenos Aires. Series 1, 1879–1911. Monthly. Series 2, 1926–1930.

Indexes: 1879–1895, in 17 (1896), 262–291; 1896–1905, in 23 (1909), 299–312.   [607]

Revista de la Sociedad Geográfica Argentina
Buenos Aires, 1881–1890.   [608]

Atlas

Biedma, J. J., and Cárlos Beyer. *Atlas histórico de la República Argentina.* Buenos Aires, Estrada, 1909. 19 pl.; 56 pp. 30 x 37 cm.   [609]
Hist. of discovery and exploration, administrative boundaries, military campaigns, etc., of the period 1492–1870.

EUROPE (EXCLUDING U.S.S.R.)

(See also [59], regional geography; [88, 90], guidebooks; [98], topographic maps; [160], geographical studies; [189], geographical education.)

Geomorphology: Spreitzer [214], 1938, pp. 217–243.—Soils: Giesecke [230], 1939, pp. 226–250.—Economic geography: Lütgens [284], 1935, pp. 246–318; 1936, pp. 3–85.—Political geography: Vogel [318], 1934, pp. 195–269.   [610]
Review of 1 English and 3 American college textbooks on the geography of Europe, in *Geog. Rev.* [476], 26 (1936), 518–520.   [611]
In 1942 the Census Library Project of the U.S. Library of Congress, Reference Department, and the Bureau of the Census, issued a preliminary list for the immediate use of the War Agencies of the U.S. Government entitled *Recent Censuses in European Countries.* In 1945 it issued two annotated preliminary mimeographed bibliographies: *National Censuses and Vital Statistics in France Between Two World Wars, 1921–1942,* 22 pp., and *National Censuses and Official Statistics in Italy Since the First World War, 1921–1944,* 58 pp.   [612]

Philippson, Alfred. *Europa (Dritte Auflage)*: *Europa ausser Deutschland.* Leipzig, Bibliographisches Institut A.G., 1928. (Sievers, *Allgemeine Länderkunde* [55].)   [613]
Selected bibl. on pp. 549–557; place cl.

The Blue Guides. London, Macmillan; Ernest Benn. *Les Guides Bleues,* Paris, Hachette.   [614]
Recent guides cover: England (1939), Wales (1936), Scotland (1932),

Great Britain (1932), London and environs (1930, 1935, 1938), Paris and environs (1927), Southern France (1926), French Alps (1926), Normandy (1928), Brittany (1928), Northwestern France (1932), Northeastern France (1930), Belgium (1929), Holland and the Rhine (1933), Switzerland (1930), Northern Italy (1937), Southern Italy (1929), Southern Spain and Portugal (1929), Northern Spain (1930).

Publication of these guides, begun immediately after World War I under the editorship of Findley Muirhead, was continued by his son, L. R. Muirhead. They are also known as the Muirhead Guides. The earlier issues were published by Macmillan in close association with the French series of *Guides bleues,* which in turn were an outgrowth of the *Guides Joanne,* an older series of local French guidebooks.

Maps and Atlases

American Geographical Society. *Environment and Conflict in Europe: Eighteen Basic Maps.* 2d ed., rev. New York, 1940, 30 pp. (*Amer. Geog. Soc. Special Publ., No. 24.*)   [615]
The list of "source materials" used in the preparation of the maps (printed on one large sheet) which the brochure accompanies is a summary of map sources. Refs. are included for economic, political, and linguistic material.

Koch, W., and C. Opitz, founders. *Eisenbahn- und Verkehrs-Atlas von Europa.* 13th ed. Leipzig, Arnd, 1929–1930. 89 pl.; 173 pp. (mostly index). 22 x 34 cm.   [616]
Sectional maps on scales of 1:600,000 and smaller showing: (1) railroads cl. according to systems, functions (through, express, local), gauge, and number of tracks; (2) stations, also cl. index to stations. An elaborate key for determining distances between stations in Germany is bound in same volume (194 pp.).

Poole, R. L. *Historical Atlas of Modern Europe, from the Decline of the Roman Empire: Comprising also Maps of Parts of Asia, and of the New World Connected with European History.* Oxford, Clarendon Press, 1902. 90 pl., each acc. by one or more pages of text. 31 x 39 cm.   [617]
Largely based on Spruner and Mencke [141], 1880.

Northwestern Europe and Finland

Braun, Gustav. "Landeskunde von Nordeuropa (im engeren Sinn): Norwegen, Schweden, Finnland," *Geog. Jhrb.* [6], 44 (1929), 225–251.   [618]

Covers Norway, 1911–1928; Sweden, 1914–1928; Finland, 1919–1928. Place cl.; 337 entries; ann.; auth. index.—See also [1157–1159, 1162].

THE BRITISH ISLES

(See also [160], geographical studies.)

Mudge-Winchell [1]: *Great Britain:* almanacs, general, a 121–122; bibliographies, national and trade, a 384–390, b 54–55, c 77, d 71–73; bibliographies, historical, a 355, b 51, c 70–71, d 65; commercial handbook (1930), a 150; dissertations, a 30–31; encyclopedias, a 43–44; gazetteers, a 324; geographical names, a 329–330, b 46–47, c 66, d 60–61; government documents, a 370–372, c 73–74; historical dictionaries, a 353, b 50; histories, a 353–355, b 50, c 70, libraries, catalogues, a 423–425, b 57, c 81, d 79; libraries, special collections, a 422; periodicals, bibliographies of, a 19–20, b 10–11; periodicals, indexes to, a 6–8, b 9, c 9, d 7; societies, a 35–36, c 14; statistical abstracts, yearbook, etc., a 123–124. *Ireland:* bibliographies, historical, b 51, c 71, d 66; commercial handbook (1928), a 150; geographical names, a 330–331; government documents, a 373; statistical abstracts, a 126.　　　　　　　　　　　　　　　　　　　　　　　　　[619]

Darbishire, B. V. "Die Britischen Inseln (1929–34)," *Geog. Jhrb.* [6], 50 (1935), 3–46.　　　　　　　　　　　　　　　　　　　　　　　　[620]
By topics and by regions, subcl. by topics; 476 entries; ann.; auth. index.

Humphreys, A. L. *A Handbook to County Bibliography, Being a Bibliography of Bibliographies Relating to the Counties and Towns of Great Britain and Ireland.* London, Strangeways, 1917. 501 pp.　　　　　　　　　　　　　　　　　　　　　　　　　　　　[621]

———

Mitchell, Sir Arthur, and C. G. Cash. *A Contribution to the Bibliography of Scottish Topography.* Edinburgh, 1917. Vol. 1, 459 pp.; Vol. 2, vi + pp. 449–706. (*Publications of the Scottish History Society,* 2d series, Vols. 14 and 15.)　　　　　　　　　　[622]
"A well-nigh exhaustive catalogue of books, papers, maps, etc., dealing with Scottish topography in its widest sense" (Suppl. No. 1 to the *Geographical Journal* [625], June, 1918, p. 8). Topical and place cl.

Black, G. F. *A List of Works Relating to Scotland.* (Reprinted, with additions, from *Bull. of the New York Public Library,* Vol. 18,

1914.) New York, New York Public Library, 1916. 1,241 pp. [623]
Incl. numerous refs. to period. arts. and monographs in society publs.
By topics, incl. bibls., periods., and transactions, public documents,
anthropology, hist. and description, economics, geology, geog., physical
geog.; auth. index.

———

"List of Works in the New York Public Library Relating to Ireland,
the Irish Language and Literature, etc.," *Bull. of the New York
Public Library,* 9, (1905), 90–104, 124–144, 159–184, 201–229, 249–
280.   [624]
By topics, incl. bibls., periods., public docs. and official publs., census
and vital statistics, hist., topography, maps, local hist.

———

Geographical Periodicals and Series

(See also [43, 47], general series.)

The Geographical Journal
   London, Royal Geographical Society. 1893–. Monthly. Succeeded
Journal of the R.G.S., 1830–1880; *Proceedings,* 1855–1878; *Proceed-
ings,* New Series 1879–1892. Indexes: *Journal of the R.G.S.,* 5 in-
dexes, one for each successive 10 vols., 1830–1880. See also "List of
Papers" and "List of Maps and Other Illustrations Contained in the
Journals and Proceedings of the Royal Geographical Society" in
C. R. Markham, *The Fifty Years' Work of the Royal Geographical
Society* (London, 1881), pp. 153–242. A *General Index to the Proceedings, 1855–1878* was published in 1920; to the *Proceedings,* New
Series, in 1896. Four *General Indexes to the Geographical Journal*
have been published: (1) for 1893–1902 in 1906, (2) 1903–1912 in
1925, (3) 1913–1922 in 1930, and (4) 1923–1932 in 1935.   [625]
   The files of the R.G.S. publs. are of particular value for the light
which they throw not only on general geog. investigation, but more
especially on the progress of exploration in the less-known parts of
the world. These publications have at all times contained book reviews
and bibl. lists. See also [8].

The Scottish Geographical Magazine
   Edinburgh, The Royal Scottish Geographical Society, 1885–. Recently 3 numbers a year. Index: 1885–1934.   [626]
The Journal of the Manchester Geographical Society
   1885–. Annual (formerly quarterly or irregularly).   [627]

The Journal of the Tyneside Geographical Society
Newcastle upon Tyne. 1899–1915. New Series, 1936–[1938–39].
Irregularly.   [628]

Geography
Manchester, The Geographical Association. 1901–. Quarterly (formerly 3 numbers a year). Title to 1927 was *The Geographical Teacher*.   [629]
Devoted to geog. education and research. Book notices and reviews. Important in reflecting the development of modern geog. in Great Britain.

The Geographical Magazine
London, published for the Geographical Magazine, Ltd., by Chatto and Windus. 1935–. Monthly.   [630]
Popular illustrated magazine.

The Institute of British Geographers. *Publications*. London, 1935–[1939].   [631]
5 long papers were published. The Institute is an organization of academic geographers.

Maps and Atlases

Official maps: Thiele [97], 1938, pp. 51–58, 242–257.   [632]

Fordham, Sir H. G. *Hand-List of Catalogues and Works of Reference Relating to Carto-bibliography and Kindred Subjects for Great Britain and Ireland, 1720–1927*. Cambridge, University Press, 1928. 26 pp.   [633]
Chron.; auth. index.

Chubb, Thomas. *The Printed Maps in the Atlases of Great Britain and Ireland . . . 1579–1870*. London, The Homeland Association, 1927. 496 pp.   [634]
A monumental work based primarily on the atlas collection in the British Museum. 3 main parts relating to the atlases of England and Wales, Scotland, and Ireland, respectively; biographical notes; genl. index.

Maps for the National Plan. Prepared by the Association for Planning and Regional Reconstruction. London, Lund Humphries, n.d., [1945?]. 119 pp. incl. 15 pl.   [635]

"A background to the Barlow, Scott, and Beveridge reports." Maps in black and white illustrating fog and sunshine, chief urban areas, govt. boundaries, population, industrial conditions, employment, etc.

Messer, Malcolm. *An Agricultural Atlas of England and Wales.* 2d ed., rev. (Agricultural Economics Research Inst., University of Oxford.) Southampton, published by direction of the Ministry of Agriculture and Fisheries by the Ordnance Survey, [1932?]. 50 pl., 2 pp. text. 48 x 48 cm.   [636]
47 dot maps (1:1,500,000) printed on transparent sheets, showing mountain and heath land, permanent grass, arable land, crops, livestock, and farm labor, as of 1928. Geological, rainfall, and relief maps, supplied in a pocket, may be inserted under the transparencies for purposes of correlation.

Wood, H. J. *An Agricultural Atlas of Scotland.* London, Gill, 1931. 16 uncolored pls., 3 uncolored maps in text; 64 pp. 24 x 30 cm. [637]
Mostly crops and livestock; also moorlands, lochs, and forests, monthly rainfall, grasslands, agricultural regions.

Stamp, L. D. *An Agricultural Atlas of Ireland.* London, Gill, 1931. 33 uncolored maps and diagrams in text; 60 pp. 24 x 30 cm.   [638]
Mostly crops and livestock; also relief, annual rainfall, agricultural regions.

Bartholomew, J. G. *Survey Atlas of Scotland: A Series of Sixty-Eight Plates of Maps and Plans, with Descriptive Text, Illustrating the Topography, Physiography, Geology, Climate, and the Historical, Political and Commercial Features of the Country.* (Royal Scottish Geographical Society.) 2d ed. Edinburgh, The Edinburgh Geographical Institute, 1912. 68 pl. 31 x 46 cm.   [639]
History (pl. 1–9); relief, geology, land surface, meteorology, population, roads and railways (pl. 10–15); reduced Ordnance Survey locational maps, ½ in. to the mile, with hypsometric tints (pl. 16–60); economic map (pl. 61); town plans (pl. 62–68).

NORWAY

(See also [161], geographical studies; [1162], bibliography.)

Mudge-Winchell [1]: bibliographies, a 407–408; bibliographies, historical, a 359; encyclopedia, a 49; commercial handbook (1920), a 150; periodicals, bibliographies of, a 22; periodicals, indexes, a 11, b 9; statistical yearbooks, a 127.—Official maps: Thiele [97], 1938, pp. 280–281.   [640]

Evers, W. "Stand und Aufgaben der Geographie in Norwegen," *Zeitschrift für Erdkunde* [723], 7 (1939), 689–706.   [641]
A survey of recent geog. publs. on Norway, with primary emphasis on physical geog. By topics, incl. bibls.; 270 entries.

Periodical

Norsk Geografisk Tidsskrift
     Oslo, Norske Geografiska Selskabs. 1926–. Quarterly. Succeeded *Aarbok,* 1889–1919/21.   [642]

Atlas

Nissen, Per, ed. *Økonomisk-geografisk atlas over Norge, med en oversigt over de kulturelle og økonomiske forhold saerlig naerings-veiene* [Economic-Geographical Atlas of Norway, with a Survey of the Cultural and Economic Relations and Especially of Trade]. Christiania, H. Aschehoug, 1921. 48 pl.; 66 pp. 39 x 50 cm.   [643]
Physical geog., agriculture, forestry, mining, fishing, whaling, manufactures, commerce, population, etc. (maps and graphs).

SWEDEN

Mudge-Winchell [1]: bibliographies, a 412; bibliographies, historical, a 360; encyclopedias, a 50; bibliographies of periodicals, a 22; statistical yearbooks, a 128.—See also [1159].   [644]

The Sweden Year-Book, 1938. Uppsala, Almqvist & Wiksells, 1938.   [645]
Bibl. on pp. 336–371 of books in English on Sweden and literary works translated into English from Swedish: by topics, incl. geog., travel.

Andersson, Ernst, ed. *Bok-Katalog omfattande geografi och resor (Sverige), åren 1901–1928.* Stockholm, Utgivarens Förlag, 1931. 134 pp.   [646]
Books, cl. by provinces with a section on Sweden as a whole; auth. index.

Periodicals and Series

Ymer
     Stockholm, Svenska Sällskapet för Antropologi och Geografi. 1881–. Quarterly. Indexes: 1881–1910; 1911–1925.   [647]

Occasional arts. and reviews in English and German as well as in Swedish.

Geografiska Annaler
Stockholm, Svenska Sällskapet för Antropologi och Geografi. 1919-. Quarterly. [648]
Aims to be an international organ of scientific geog. Arts. and book reviews in English, French, and German.

Globen: Meddelanden utgivne av Generalstabens Litografiska Anstalt
Stockholm. 1922–[1939]. 8 nos. a year. Index: 1922–1931. [649]
The best source for information on Swedish maps. Incidental information on mapping in other countries.

Svensk Geografisk Årsbok: The Swedish Geographical Yearbook
Lund, Sydsvenska Geografiska Sällskapet. 1925–[1940]. Index: 1925–1934. [650]
Includes a systematically arranged bibl. of geog. publs. relating to Sweden and Swedish geog. publs. relating to other parts of the world.

Jorden Runt
Stockholm. 1929–[1941]. Monthly. [651]
Well-illustrated, popular journal; though nominally universal in scope, contains a good deal of local material.

———

Geographica: Skrifter från Upsala Universitets Geografiska Institution
1936–[1941]. Irregularly. [652]
Includes monographs dealing with various phases of Swedish geog.

Gothia: Meddelanden från Geografiska Föreningen i Göteborg
Göteborg. Ny serie, 1932–[1937]. Irregularly. Preceded by *Meddelanden,* 1912, 1917, 1924, 1928. [653]
Each number contains several long articles "preferably by Göteborg geographers and papers treating greater Göteborg and its sphere of influence . . . but also concerning the geography of Europe and Trans-Oceania."

Atlas

Flach, Wilhelm, H. J. Dannfelt, and Gustav Sundbärg. *Sveriges jordbruk, vid 1900 talets början statistiskt kartverk: L'Agriculture*

en Suède au commencement du XXème siècle. Göteborg, 1909. 89 colored pl.; 262 pp. 30 x 49 cm.   [654]
The main emphasis is on rural economy, with sections on geology, meteorology, and demography. Table of contents and glossary in French.

DENMARK

Mudge-Winchell [1]: bibliographies, a 396–397, b 55, c 77; bibliographies, historical, a 350; encyclopedias, a 45–46; periodicals, indexes, a 9–10; statistical yearbook, a 125.—See also [1157].   [655]

Nielsen, Niels. "Dänemark und die Färöer (1914–30)," *Geog. Jhrb.* [6], 46 (1931), 203–219.   [656]
By topics, with a section on regions; 252 entries; ann.; auth. index.

Periodical

Geografisk Tidsskrift
    Copenhagen, Kongelige Danske Geografiske Selskab. 1877–. Annual since 1936 (previously monthly). Indexes: 1877–1910, 1911–1930.   [657]

FINLAND

Mudge-Winchell [1]: bibliographies, a 398–399, b 55; statistical yearbooks, a 125, c 26.—See also [1158].   [658]

Neuvonen, E. K. "A Short Bibliography on Finland: A Selection of Recently Published Works Concerning Finland in English, French, Italian and Dutch . . . ," *Finland Year Book,* 1939/40, Helsinki, [1939], pp. 394–406.   [659]
Topical cl.

Van Cleef, Eugene. *Finland—The Republic Farthest North: The Response of Finnish Life to Its Geographic Environment.* Columbus, Ohio Univ. Press, 1929.   [660]
Bibl. on pp. 205–209 of selected titles of works in English, German, and French; alph. by auths.; 74 entries.

Periodicals and Series

Fennia
    Helsinki, Societas Geographica Fenniae (Geografiska Sällskapet i

Finland). 1889–. Prior to 1924 subtitle was *Bulletin de la Société de Géographie de Finlande;* prior to 1921 published by Sällskapet for Finlands Geografi (see note on [662]). [661]
Arts. in Swedish, Finnish, German, French, English.

Terra: Suomen Maantieteellisen Seuran Aikakauskirja: Geografiska Sällskapets i Finland Tidskrift [Terra: Journal of the Geographical Society of Finland]
    Helsinki, 1913–. Quarterly (formerly annual). Succeeds the annual *Geografiska Föreningens Tidskrift,* 1891–1912. The Geografiska Föreningen i Finland was merged with the Sällskapet for Finlands Geografi in 1921 to form the Geografiska Sällskapet i Finland. [662]
Arts. in Finnish and Swedish, occasionally with French, German, or English résumés.

––––

Publicationes Geographici Universitatis Turküensis [*Aboensis,* prior to 1936]. Helsinki, later Turkü. 1927–. Irregularly. [663]
Acta Geographica. Helsinki, Societas Geographica Fenniae (Geografiska Sällskapet i Finland). 1925–. Irregularly. [664]
Monographs (mostly in German, a few in English or Swedish) dealing with "general or methodical investigations or researches on foreign countries."

Meddelanden af Geografiska Föreningen. Helsingfors, Geografiska Föreningen i Finland. 1892–1920. Irregularly. Vols. 1–3, 1892–1896, entitled *Vetenskapliga Meddelanden,* etc. Vol. 11 for 1915–1920 published in 1920. [665]
In Finnish, Swedish, German, and French. Summaries of Finnish and Swedish arts. in German or French.

––––

Atlas

Atlas of Finland, 1925. (The Geographical Society of Finland.) Helsinki, 1925–1928. 38 pl. 33 x 46 cm. In Finnish, Swedish, and English. Accompanied by volume of text (English ed., Helsinki, 1929, 320 pp.). [666]
A work somewhat comparable to the *Atlas of Czechoslovakia* [800], covering a very wide variety of topics concerning the physical, economic, and social geog. of Finland. See *Geog. Rev.* [476], 20 (1930), 350.

THE NETHERLANDS

Mudge-Winchell [1]: atlas, historical, a 337; bibliographies, a 397–398, b 55, c 77; bibliographies, historical, a 358, b 52, 55; dissertations, a 31; three encyclopedias, a 46, b 13, c 16; government documents, a 374; libraries, a 421; periodicals, indexes, a 10; statistical yearbook, a 127.—Official maps: Thiele [97], 1938, 278–279.   [667]

Oestreich, K. "Die Niederlande (1928–38)," *Geog. Jhrb.* [6], 54, Pt. 2 (1939), 597–634.   [668]
By topics, with a section on regional monographs; 293 entries; ann.; auth. index.

Van der Meulen, R., and others. *Algemeene aardijkskundige bibliographie van Nederland.* Uitgegeven door de Afdeeling "Nederland" van het Nederlandsch Aardrijkskundig Genootschap. Leiden, Brill, 1888–1889. 3 parts: 285, 259, and 260 pp.   [669]
Part 1, Netherlands in genl. (topical cl.); regional descriptions (place cl.); Part 2, natural environment; Part 3, population and economic conditions. Parts 2 and 3 cl. by topics. Auth. and subj. indexes to each part.

Periodicals and Series

Tijdschrift van het Koninklijk Nederlandsch Aardrijkskundig Genootschap
   Amsterdam. 1874–. Bimonthly. Indexes: topical index *Systematisch Register*), 1876–1904, 1905–1922, 1923–1927, 1928–1933; alph. index, 1905–1909.   [670]
Though this journal is devoted to geog. in general, more attention is given to the Netherlands and Netherlands Indies than to other regions.

Tijdschrift voor Economische Geographie
   Amsterdam (formerly The Hague), Nederlandsche Vereeniging voor Economische Geographie. 1910–. Monthly. Index: 1910–1924.   [671]
Tijdschrift voor het Onderwijs in de Aardrijkskunde
   Vondellaan, Aerdenhout, Overveen. 1923–[1942]. Monthly; bimonthly before 1930.   [672]

———

Geographisch en Mineralogisch-Geologisch Instituut der Rijksuniversiteit te Utrecht: Geographische en Geologische Mededeelingen.

Anthropo-geographische Reeks. 1929–[1932]. Irregularly. *Physio-graphisch-geologische Reeks.* 1927–. Irregularly. [673]

Atlas

Geschiedkundige Atlas van Nederland. . . . (Commissie voor den Geschiedkundigen Atlas van Nederland.) The Hague, Nijhoff, 1913–[1938]. 19 maps (*kaarten*), many subdivided into separate sheets (*bladen*), making a total of 154 pl. 37 x 50 cm. 42 vols. of text (16 x 25 cm.) [674]
Covers history of the Netherlands from prehistoric times to the present. *Kaart* 19 in 33 sheets deals with the Netherlands colonies.

BELGIUM AND LUXEMBOURG

(See also [161], geographical studies.)

Mudge-Winchell [1]: atlas, historical, a 337; bibliographies, a 392–393; bibliographies, historical, a 348; dissertations, b 12; encyclopedia, c 69; gazetteers, a 323; geographical names, c 66; periodicals, indexes, a 9; statistical yearbook, a 124. [675]

Leyden, Friedrich. "Belgien (1915–1928)," *Geog. Jhrb.* [6], 44 (1929), 166–194. [676]
By topics, some subcl. areally; 366 entries; ann.; auth. index.

Schmithüsen, Josef. *Das Luxemburger Land.* Leipzig, S. Hirzel, 1940. (*Forschungen zur deutschen Landeskunde,* Vol. 34.) [677]
Bibl. on pp. 397–425: by topics, incl. bibls., maps, various aspects of physical and human geog.; *ca.* 550 entries.

Leyden, Friedrich. "Luxemburg (1915–28)," *Geog. Jhrb.* [6], 44 (1929), 195–196. [678]

Periodicals and Series

Bulletin de la Société Royale de Géographie d'Anvers
Antwerp. 1876–[1940]. Quarterly. Publication was suspended 1914–1919. Index: 1876–1926. [679]
Mémoires of the same society were published in 1879, 1883, 1886, and 1895.

Bulletin de la Société Royale Belge de Géographie
Brussels. 1877–[1943]. Quarterly. The period 1914–1919 was cov-

ered by one fascicule published in 1920. Index: 1876–1901.   [680]
Book reviews and lists of accessions to the library. Valuable for Belgian
Congo.

Bulletin de la Société Belge d'Études Géographiques
   Louvain, Institut Géographique de l'Université. 1931–. Semiannual (annual 1939–1945).   [681]
Monographs and book reviews. See *Geog. Rev.* [476], 36 (1946), 155–
156.

*Le Mouvement géographique: Journal populaire des sciences géo-
graphiques*
Brussels. 1884–1922. Weekly.   [682]
One of the few geog. weeklies. Besides geog. news of interest it contains
brief notices and reviews of books.

———

*Cercle des Géographes Liégeois: Travaux, et Travaux du Séminaire
de Géographie de l'Université de Liége.* Liége. 1929–. Irregularly.   [683]
Deals largely with Belgium. Most of the material is reprinted from
other journals.

———

Atlas

Essen, L. van der, ed. *Atlas de géographie historique de la Belgique.*
. . . Brussels, Van Oest, 1919–1932. 6 of 12 contemplated pls. have
appeared. Each pl. accompanied by some 15 pp. of text. 25 x 33
cm.   [684]
The maps which have appeared cover Belgian hist. at the end of the
11th century and from 1648 to 1839.

FRANCE

(See also [160, 161], geographical studies; [1163], bibliography.)

Mudge-Winchell [1]: almanacs, general, a 122; bibliographies, national and trade, a 399–401, b 55–56, c 78; bibliographies, historical,
351–352; commercial handbook (1931), a 150; dissertations, a 31;
encyclopedias, a 46–47, b 13, c 16; gazetteers, a 324; geographic
names, a 328; historical dictionaries, a 350–351; libraries, a 421;
libraries, catalogues, a 425, b 57, c 81; periodicals, bibliographies,
a 21, c 11; periodicals, indexes, a 10; societies, a 36, c 14.   [685]

Conover, Helen F. *France: A List of References on Contemporary Economic, Social and Political Conditions.* (The Library of Congress, General Reference and Bibliography Division.) Washington, 1944. 183 mimeo. pp.   [686]
"The bibliography, while primarily concerned with economic life, provides a brief coverage of the related fields conditioning and conditioned by it." By topics, incl. genl. ref. works, economic studies, social conditions, political conditions, physical aspects (incl. atlases and maps, physical geog., geology, climate, guidebooks); 1,366 entries; auth., period., and subj. indexes.

Ormsby, Hilda. *France: A Regional and Economic Geography.* New York, Dutton, [1931]. 530 pp.   [687]
Each regional chapter is followed by a list of refs. to books and period. arts.; also a short bibl. covering genl. works, agriculture, industry, and communications (pp. 495–499).

Almeida, P. Camena d'. "Frankreich (1916–27)," *Geog. Jhrb.* [6], 43 (1928), 276–312.   [688]
By topics, incl. a section "Chorographie" (regional studies); 725 entries; ann.; auth. index.

Martonne, Emmanuel de. *Les Regions géographiques de la France.* Paris, Flammarion, 1921. 190 pp.   [689]
Selected refs. to regional studies by professional geographers are given at the ends of the chapters (not included in the English translation, London, 1933).

The Principal Geographical Periodicals

(See also [48] commercially published series.)

Comptes rendus des Congrès Nationaux des Sociétés Françaises de Géographie
1878–1914.   [690]
The following list shows places and dates of the Congresses and, in parentheses, the dates of publ. of the *Comptes rendus.* The place of publ. is the same as that of the Congress except where otherwise indicated.

First, Paris, 1878 (1879); Second, Montpellier, 1879 (1880); Third, Nancy, 1880 (1880); Fourth, Lyon, 1881 (1882); Fifth, Bordeaux, 1882 (1883); Sixth, Douai, 1883 (1883); Seventh, Toulouse, 1884; Eighth, Nantes, 1886; Ninth, Le Havre, 1887 (1887); Tenth, Bourg, 1888 (1889); Eleventh, Montpellier, 1890 (1891); Twelfth, Rochefort-sur-mer, 1891 (1893); Thirteenth, Lille, 1892 (Lille, n.d.); Fourteenth, Tours, 1893 (see *Rev. Soc. de Géog. de Tours,* 1893, 89–102); Fifteenth, Lyon, 1894 (1895); Sixteenth, Bordeaux, 1895 (1896); Seventeenth, Lorient, 1896 (1897); Eighteenth, St. Nazaire, 1897 (1898); Nineteenth, Marseille, 1898 (1899); Twentieth, Algiers, 1899 (1900); Twenty-first, Paris, 1900 (1901); Twenty-second, Nancy, 1901 (1902); Twenty-third, Oran, 1902 (1903); Twenty-fourth, Rouen, 1903 (1904); Twenty-fifth, Tunis, 1904 (1904); Twenty-sixth, St. Etienne, 1905 (1906); Twenty-seventh, Dunkirk, 1906 (1907); Twenty-eighth, Bordeaux, 1907 (1908); Twenty-ninth, Nancy, 1909 (see *Bull. Soc. de Géog. de l'Est,* 30 (1909), 298–308); Thirtieth, Roubaix, 1911 (Lille, 1912); Thirty-first, Paris, 1913 (1914); Thirty-second, Brive, 1914 (Brive, n.d.).

La Géographie

Paris, Société de Géographie. 1900–1939. Normally monthly. In 1931–1933 entitled *Terre, air, mer: La Géographie.* Merged with *Annales de géographie* [694] in 1941. *La Géographie* succeeded: (1) *Bulletin de la Société de Géographie,* which appeared in 7 series from 1822 to 1899, and (2) *Compte rendu des séances de la Société de Géographie et de la Commission Centrale,* 1882–1899. Indexes (*Tables générales analytiques*): *Bulletin,* first and second series, 1822–1843; third and fourth series, 1844–1860; fifth, sixth, and seventh series, 1861–1899; *Comptes rendus,* 1882–1899; *La Géographie,* 1900–1939.   [691]
In 1919 an extensive bibl. was begun and continued in each issue until 1931. Titles of books and period. arts. are given, and critical notes are appended to the more important items.

Revue economique française

Paris, Société de Géographie Commerciale et d'Études Économiques. 1919–[1939]. Irregularly. Succeeds *Bull. de la Société de Géographie Commerciale de Paris,* 1878–1918. Name of the society changed to present name in 1932.   [692]

Bulletin de la Section de Géographie

Paris, Section de Géographie du Comité des Travaux Historiques et Scientifiques, Ministère de l'Instruction Publique et des Beaux Arts. 1913–1939. Annual (quarterly before 1915). Succeeded *Bulletin de géographie historique et descriptive,* 1886–1912.   [693]
Book reviews. Auth. index of memoirs published in first 20 vols. (1886–1905) in 20 (1905), 475–490.

Annales de géographie: Bulletin de la Société de Géographie
 Paris, Armand Colin. 1891–. Bimonthly. Subtitle added in 1941
 (see [691]). Indexes (*Tables décennales*): 1891–1901, 1902–1911,
 1912–1921, 1922–1931.   [694]
 A scholarly and scientific periodical of the highest order. For the bibliography see [7]. See also *Geog. Rev.* [476], 33 (1943), 330–331.
 Topical and regional cl. (system.); auth. indexes.

Revue de géographie annuelle
 Paris, Librarie Charles Delagrave. 1906–1924. Annual (Vol. 9
 covers 1916–1921). Succeeded the monthly *Revue de géographie*,
 1877–1905. Indexes: 1877–1883, 1884–1894.   [695]
 Monographs. After 1906 dealt more especially with physical geog.

Other Geographical Periodicals

Bulletin de la Société de Géographie de Lyon
 1875–1914, 1921–. Annual (monthly prior to 1915). Merged with
 Les Études rhodaniennes [701], in 1942 (?).   [696]
Bulletin de la Société Languedocienne de Géographie
 Montpellier. 1878–[1940]. Quarterly. Indexes: 1878–1887 (in Vol.
 11, 1888), 1888–1897.   [697]
Publications de la Société de Géographie de Lille
 1882–. Irregularly. Prior to 1941 entitled *Bulletin*.   [698]

Bulletin de l'Association de Géographes Français
 Paris. 1924–. Monthly (except Aug., Sept., Oct.).   [699]
 The Association de Géographes Français is an organization of pro-
 fessional geographers, akin to the Association of American Geographers.

Revue Géographique des Pyrenées et du Sud-Ouest
 Institut de Géographie, Faculté des Lettres, Université de Tou-
 louse. 1930–. Quarterly.   [700]
 In the Jan. number, 1931, publ. was begun of an annual bibl. of the
 French Pyrenees which covers the years 1929–1936.

*Les Études rhodaniennes: Revue de géographie régionale: Bulletin de
 la Société de Géographie de Lyon et de la Région Lyonnaise*
 Institut des Études Rhodaniennes de l'Université de Lyon. 1925–.
 Second subtitle added 1942 (?) (see [696].) 4 numbers a year,
 usually published as 3. Index: 1925–1934, in Vol. 10, 1934.   [701]
 Long scholarly arts. and some reviews. Earlier issues contained annual
 bibls.

L'Information géographique
Paris, Baillière, 1936–[1941]. 5 or 6 nos. a year.   [702]
French equivalent of the *Journal of Geography* [478].

A travers le monde
Société de Géographie de Compiègne. 1946–. Monthly.   [702a]

Bulletin de la Société de Géographie de Marseille
1877–[1938]. Annual. Index: 1877–1905 in Vol. 30, 1906.   [703]
Book reviews and an extensive bibl. of acquisitions to the Library.

Bulletin de la Société de Géographie de Rochefort
1878–[1930]. Irregularly (formerly quarterly).   [704]
Devoted to "agriculture, letters, science, and art" as well as to geog.
Relatively unimportant except for colonial and naval hist.

*Revue française de l'étranger et des colonies: Exploration et gazette
géographique*
1885–1914. Monthly.   [705]

Atlas

Atlas de France. (Comité National de Géographie.) Paris, Éditions
Géographiques de France, 1933–. *ca.* 80 pl. 32 x 50 cm.   [706]
The original plan for this outstanding national atlas has been modified
in minor details. It called for 11 plates for geomorphology and geophysics, 8 for climatology, 6 for hydrography, 8 for biogeog., 9 for
agriculture, 11 for industry, 11 for commerce, and 16 for "human and
political geog." (chorography, administration, political and religious
opinions, population density and movements, etc.). See reviews in
Geog. Rev. [476], 24 (1934), 678–680; 28 (1938), 706.

Central Europe

Krebs, Norbert, ed. *Atlas des deutschen Lebensraumes in Mitteleuropa.* (Preussische Akademie der Wissenschaft.) Leipzig, Bibliographisches Institut, 1937–. Pl. with text. Issued in sections for loose-
leaf binder. 34 x 47 cm.   [707]
Designed when completed to cover a wide variety of topics in the fields
of physical, economic, human, and historical geog. Copy in Library of
Amer. Geog. Soc. contains handsome colored plates illustrating relief,
soils, temperature, rainfall, natural vegetation, forests, agricultural land
utilization, population (distribution and trends), birthrates, historical
geog.

(See also [161, 166], geographical studies.)

Mudge-Winchell [1]: almanacs, general, a 122; bibliographies (national, trade, etc.), a 402–403, b 56, c 78, d 65; bibliographies, historical, a 352–353, b 50, c 70; dissertations, a 31–32; encyclopedias a 47–48, b 13; gazetteers, a 324; government documents, a 374, d 70; libraries, a 421; libraries, catalogues, a 425, b 57, c 81; periodicals, bibliographies of, a 21–22; periodicals, indexes, a 10; societies, a 36; statistical yearbook, a 126.—Soils: Giesecke [230], 1939, pp. 206–226.—Economic geography: Lütgens [284], 1935, pp. 251–305.—Political geography: Vogel [318], 1934, pp. 195–223.—Geographical science in German universities and institutes: Tulippe [166], 1930.

[708]

Krebs, Norbert, ed. *Landeskunde von Deutschland*. Leipzig und Berlin, B. G. Teubner, 1931 [–1935]. 3 [?] vols. [709]
Each volume contains selected bibls. (topical and place cl.): (1) Hans Schrepfer, "Der Nordwesten," 1935, pp. 258–279 (722 entries); (2) Bernhard Brandt, "Der Nordosten," 1931, pp. 139–148 (305 entries); (3) Norbert Krebs, "Der Südwesten," 2 Aufl., 1931, pp. 203–219 (567 entries).

Geisler, Walter. "Deutsches Reich": (a) "Gesamtgebiet (1927–32)," *Geog. Jhrb.* [6], 49 (1934), 62–78; (b) "Norddeutschland und Mitteldeutschland (1927–32)," *ibid.*, 49 (1934), 3–62. [710]
(a) Topical cl.; 229 entries; (b) by regions, a few with topical subcl.; 910 entries. Ann.; auth. index. See also [712].

Braun, Gustav. *Deutschland, dargestellt auf Grund eigener Beobachtung, der Karten und der Literatur*. 2d ed. Berlin, Borntraeger, 1926–1934. 3 parts in 4. [711]
Divided into chapters on the larger regions of Germany, each followed by a selected bibl. subcl. by districts.

Berninger, O. "Deutsches Reich: Süddeutschland und Rheingebiet (1927–32), *Geog. Jhrb.* [6], 48 (1933), 159–243. [712]
By regions, a few with topical subcl.; 992 entries; ann.; auth. index. See also [710].

Bericht über die neuere Litteratur zur deutschen Landeskunde. Herausgeg. im Auftrage der Zentralkommission für wissenschaftliche Landeskunde von Deutschland. Vol. 1 covering 1896–1899,

Berlin, 1901, 259 pp.; Vols. 2 and 3 covering 1900–1903, Breslau, 1904, 421 pp.; 1906, 256 pp.   [713]
By topics, a few subcl. by places; fully ann.; auth. index in each vol.

Richter, P. E. *Bibliotheca geographica Germaniae: Litteratur der Landes- und Volkskunde des Deutschen Reichs.* Bearbeitet im Auftrage der Zentral-Kommission für wissenschaftliche Landeskunde von Deutschland. Leipzig, Engelmann, 1896. 851 pp. Autoren-Register, 1897, 54 pp.   [714]
Many items are included the relation of which to geog. is but distant, and publs. in languages other than German on the whole have been neglected. By topics, incl. (1) bibls.; (2) surveys, maps, and plans; (3) regional studies and works of travel; (4) physical and biological geog.; (5) inhabitants; (2) and (3) are subcl. by places; auth. index.

The Principal Geographical Periodicals

Verhandlungen und wissenschaftliche Abhandlungen des Deutschen Geographentages
   1881–[1937].   [715]
The following list shows the places and dates of the Geographentage and, in parentheses, the dates of the publ. of the *Verhandlungen.* The first to the twenty-first inclusive were published at Berlin, the twenty-second to twenty-sixth inclusive at Breslau. The words "und Wissenschaftliche Abhandlungen" were added to the title in 1925.
   First, Berlin, 1881 (1882); Second, Halle, 1882 (1882); Third, Frankfort-on-Main, 1883 (1883); Fourth, Munich, 1884 (1884); Fifth, Hamburg, 1885 (1885); Sixth, Dresden, 1886 (1886); Seventh, Karlsruhe, 1887 (1887); Eighth, Berlin, 1889 (1889); Ninth, Vienna, 1891 (1891); Tenth, Stuttgart, 1893 (1893); Eleventh, Bremen, 1895 (1896); Twelfth, Jena, 1897 (1897); Thirteenth, Breslau, 1901 (1901); Fourteenth, Cologne, 1903 (1903); Fifteenth, Danzig, 1905 (1905); Sixteenth, Nuremberg, 1907 (1907); Seventeenth, Lübeck, 1909 (1910); Eighteenth, Innsbruck, 1912 (1912); Nineteenth, Strassburg, 1914 (1915); Twentieth, Leipzig, 1921 (1922); Twenty-first, Breslau, 1925 (1926); Twenty-second, Karlsruhe, 1927 (1928); Twenty-third, Magdeburg, 1929 (1930); Twenty-fourth, Danzig, 1931 (1932); Twenty-fifth, Bad Neuheim, 1934 (1935); Twenty-sixth, Jena, 1936 (1937).

Zeitschrift der Gesellschaft für Erdkunde zu Berlin
   1902–[1944]. 10 numbers a year. Succeeds: *Monatsberichte über die Verhandlungen der Gesellschaft für Erdkunde zu Berlin,* 1839–1853; *Zeitschrift für allgemeine Erdkunde,* 1853–1865; and the

parallel publs. *Zeitschrift der Gesellschaft für Erdkunde zu Berlin,* 1866–1901, and *Verhandlungen der Gesellschaft für Erdkunde zu Berlin,* 1873–1901. Indexes: *Übersicht der Aufsätze, Miscellen und Karten* to the *Monatsberichte* and *Zeitschrift für allgemeine Erdkunde,* 1840–1863; *Inhaltsverzeichnis* to the *Zeitschrift für allgemeine Erdkunde, Zeitschrift,* and *Verhandlungen,* 1863–1901. [716]
From 1853 through 1890 an extensive summary of geog. literature was published in the *Zeitschrift.* In 1895 this was continued in *Bibliotheca Geographica* [10].

Petermanns geographische Mitteilungen
Gotha, Justus Perthes. 1855–[1945]. Monthly. Combined with *Das Ausland* (1828–1893), *Aus allen Weltteilen* (1870–1898), and *Globus* (1862–1910). Prior to 1938 title was *Dr. A. Petermanns Mitteilungen aus Justus Perthes' Geographischer Anstalt.* Indexes (*Inhaltsverzeichnis*): 1855–1864; 1865–1874; 1875–1884; 1885–1894; 1895–1904; 1905–1934. [717]
Petermanns Mitteilungen is a bibl. repertory of the first rank (see [11]), of especial value with regard to maps and current German publs. The indexes are divided into 4 parts: (1) maps, pictures, etc. (key maps are given in the index vols. showing the areas covered by maps published in the *Mitteilungen*); (2) articles, lesser notes, and monthly record; (3) "Literaturberichte," i.e. book reviews and bibls.; (4) auth. index. Each of the first 3 of these parts is arranged in sections devoted to general topics, special topics, and regions.

Ergänzungshefte zu Petermanns geographische Mitteilungen
Gotha, Justus Perthes. 1860–[1944]. Through 1939, 238 numbers had appeared. Prior to 1938 title was *Ergänzungshefte zu Dr. A. Petermanns Mitteilungen aus Justus Perthes' Geographischer Anstalt.* [718]
Monographs and maps.

Geographische Zeitschrift
Leipzig, B. G. Teubner. 1895–[1944]. Monthly. Ed. (1940): Prof. Heinrich Schmitthenner, Univ. of Leipzig. Indexes: 1895–1904, 1915–1924 (topical and geog. cl. (system.); auth. indexes). [719]
This scholarly periodical was founded, and edited until 1934, by Professor Alfred Hettner of the University of Heidelburg. Contains book reviews, lists of new maps, and period. arts.

Mitteilungen der Geographischen Gesellschaft in München
> Munich. 1904-[1942-43]. Annual since 1936; previously 2 numbers a year (4 a year 1911–1913, 3 in 1914). Succeeds *Jaresbericht,* 1871–1902. [720]
> Monographs, book reviews, and maps of first quality. A list arranged chron. in order of vols. of the contents of the *Jahresberichte* and *Mitteilungen* to 1913 will be found in *Katalog der Bibliothek der Geographischen Gesellschaft in München* (Munich, 1914), pp. 194–204.

Geographischer Anzeiger
> Gotha, Justus Perthes. 1900-[1944]. Bimonthly (formerly monthly). Combined with the *Zeitschrift für Schulgeographie* in 1911. Prior to 1935 carried subtitle *Blätter für den geographischen Unterricht.* [721]
> Contains book reviews and abstracts.

Zeitschrift für Geopolitik
> Heidelberg (formerly Berlin), Kurt Vowinkel. 1924-[1944]. Monthly. [722]
> Ed. by Dr. Karl Hausofer of Munich, the originator of German "geopolitics'" (see Kiss [316], 1942, pp. 640–645).

Zeitschrift für Erdkunde
> Frankfurt-am-Main. 1936-[1944]. Monthly. Succeeds *Geographische Wochenschrift,* Halle, 1933–1935, weekly. [723]
> Contains book reviews.

Deutsches Archiv für Landes- u. Volksforschung
> Leipzig. 1937-[1940]. Quarterly. [724]
> Long arts. with many maps. Most numbers include a rather exhaustive bibl. of some region or country.

Geographen-Kalender
> Gotha, Justus Perthes. 1903–1914. Annual. [725]
> Record of current geog. news, addresses of geographers, geog. societies, etc.

Other Geographical Periodicals

Mitteilungen der Gesellschaft für Erdkunde zu Leipzig
> 1872-[1937/39]. Irregularly (1 vol. for two or three years); formerly annual. *Jahresberichte* from 1861 through 1871. Subsequently the *Jahresberichte* and *Mitteilungen* were published together.

Prior to 1911 the name of the society was Verein für Erdkunde zu Leipzig. [726]

Mitteilungen der Geographischen Gesellschaft in Hamburg
   1876–[1944]. Recent vols. annual. Succeeds *Jahresbericht*, 1873–75. Index: 1873–1917; *Inhaltsverzeichnis* to Vols. 1–14 (1876–1898) in Vol. 14; to Vols. 15–35 (1899–1923) in Vols. 36 (1924). [727] Arts. and maps. No book reviews or bibls.

Deutsche geographische Blätter
   Geographische Gesellschaft in Bremen. 1877–[1940]. Quarterly. Index to Vols. 1–30 (1877–1907). [728] Arts. and maps. Brief section of book reviews.

Mitteilungen der Geographischen Gesellschaft zu Rostock i. M.
   1910–[1939]. Irregularly. See also Beihefte 1–11, 1934–1938. [729]
Mitteilungen der Geographischen Gesellschaft zu Würzburg
   1925–[1939]. Annual. [730]
Frankfurter geographische Hefte
   Frankfurt-am-Main. 1927–[1939]. Normally biannual. Succeeds *Jahresbericht des Frankfurter Vereins für Geographie und Statistik,* 1836–1925. [731]
Notizblatt der Hessischen Geologischen Landesanstalt
   Darmstadt. 1854–[1938]. Annual. Prior to V. Folge, 17 Heft 1936, title was *Notizblatt des Vereins für Erdkunde und der Hessischen geologischen Landesanstalt zu Darmstadt.* [732]
Mitteilungen der Geographischen Gesellschaft und des Naturhistorischen Museums in Lübeck
   1882–[1937] (excepting 1917–1920). Annual. [733]
Mitteilungen und Jahresberichte der Geographischen Gesellschaft in Nürnberg
   1919–[1937]. Irregularly. [734]
Jahrbuch der Geographischen Gesellschaft zu Hannover
   1925–[1936/37]. *Jahresbericht,* 1878–1911; subsequently *Jahresberichte* were published in the *Jahrbuch.* [735]
Mitteilungen der Geographischen Gesellschaft (für Thüringen) zu Jena
   1882–[1936]. Indexes: 1882–1893; 1894–1907. [736]
Mitteilungen des Sächsisch-Thüringischen Vereins für Erdkunde zu Halle a. S.
   1877–[1935–36]. Annual (1 vol. for 1915–1919). [737]

Scientific arts. Extensive sections devoted to book notices and reviews. See also Beihefte 1–5, 1930–1936.

Mitteilungen des Vereins für Erdkunde zu Dresden
1905–[1935–36]. Annual since 1926; previously irregularly. Succeeds *Jahresbericht*, 1865–1901. [738]
Contains book reviews.

Jahrbuch der Pommerschen Geographischen Gesellschaft, Sitz Greifswald
1922–[1935–36]. Nominally annual, but since 1922 normally each vol. has covered two years. Succeeds *Jahresberichte*, 1882–1922. [739]
Table of contents (*Inhaltsverzeichnis*) covering 1882–1932 in *Jahrbuch*, 49–50 (1931–1932), 43–48; also list of papers publ. by the Society and dealing with Pomerania (*ibid.*, pp. 49–53).

Erde und Wirtschaft
Brunswick, Georg Westermann. 1927–1934. Quarterly. Ed. by Prof. Dr. G. Braun, Greifswald. [740]
Die neue Geographie
Brunswick, Georg Westermann. 1922–1925/26. Quarterly. Ed. by Ewald Banse. [741]
Gaea: Natur und Leben: Centralorgan zur Verbreitung naturwissenschaftlicher und geographischer Kenntnisse
Leipzig, E. H. Mayer. 1865–1909. Normally monthly. [742]
A section devoted to geog. literature.

Series

(See also [49], commercially published series.)

Geographische Abhandlungen. Stuttgart, Spemann, 1886–[1939]. Irregularly. Ed. 1886–1928 by Albrecht Penck, 1929–[1939] by Norbert Krebs. Published 1886–1900, by Hoelzel, Vienna; 1903–1928, by Teubner, Leipzig and Berlin. 3 series, 1886–1921, 1923–1928, 1929–[1939]; also Neue Folge, 1912–1917. [743]
Monographs. Especial emphasis on geomorphology.

Forschungen zur deutschen Landes- und Volkskunde. Stuttgart, Zentralkommission für wissenschaftliche Landeskunde von Deutschland. 1886–[1940]. Irregularly. [744]
Series of monographs devoted to the geog. of Germany.

Veröffentlichungen der Schlesischen Gesellschaft für Erdkunde.
Breslau. 1922–[1939]. Irregularly.   [745]
Oberrheinische geographische Abhandlungen. Freiburg, Geographisches Institut der Universität. 1926–[1939]. Prior to 1939 title was *Badische geographische Abhandlungen.*   [746]
Rhein-Mainische Forschungen. Geographisches Institut der Universität Frankfurt-am-Main. 1927–[1939]. Irregularly.   [747]
Berliner geographische Arbeiten. Geographisches Institut der Universität Berlin. 1932–[1940]. Irregularly.   [748]
Schriften des Geographischen Instituts der Universität Kiel. 1932–[1939]. Irregularly.   [749]
Wissenschaftliche Veröffentlichungen des Deutschen Museums für Länderkunde. Leipzig. N. F. 1932–[1940]. Irregularly. Succeeds *Veröffentlichungen des Städtischen Museums für Länderkunde zu Leipzig.* 1896–[1914].   [750]
Veröffentlichungen des Geographischen Instituts der Albertus-Universität zu Königsberg. 1919–[1938]. Irregularly. Reihe Geographie, 1931–[1938]; Reihe Ethnographie, 1931–1932; Ausser der Reihe, 1925–[1938].   [751]
Mitteilungen des Vereins der Geographen an der Universität Leipzig. 1911–[1937]. Irregularly.   [752]
Dresdner geographische Studien. Geographisches Institut und Geographische Arbeitsgemeinschaft, Technische Hochschule, Dresden. 1931–[1937]. Irregularly.   [753]
Wissenschaftliche Veröffentlichungen des Vereins (der Gesellschaft) für Erdkunde zu Leipzig. 1891–[1936]. Vol. 9 appeared in 1921, Vol. 10 in 1936.   [754]
Monographs.

Beiträge zur Landeskunde der Rheinlande: Veröffentlichungen des Geographischen Instituts der Universität Bonn. 1922–[1936]. Irregularly.   [755]
Monographs dealing with the German Rhine Valley.

Veröffentlichungen des Geographischen Seminars der Universität Leipzig. 1930–[1936].   [756]
Heimatkundliche Arbeiten. Geographisches Institut der Universität Erlangen. 1926–[1934]. Irregularly.   [757]
Stuttgarter geographische Studien. Stuttgart, Geographisches Seminar der Technischen Hochschule. Reihe A, 1924–[1930]; Reihe B,

Unterrichtsbeiträge, 1925–[1933]; Reihe C, *Geographische Exkursionsführer für Württemberg,* 1925 [only one number published]. Irregularly. [758]

Landeskundliche Forschungen herausgegeben von der Geographischen Gesellschaft in München. 1906–1929. Irregularly. [759]
Monographs devoted in the main to the geography of Germany.

Maps and Atlases

Official maps: Thiele [97], 1938, pp. 258–273. [760]
Keyser, Erich. "Die Landeskarten der Preussenlandes: Eine Bestandsaufnahme," *Deutsches Archiv für Landes- und Volksforschung* [724], 2, No. 2 (1938), 496–518. [761]
422 titles (chron.), ranging from 1542 to 1918.

Mitteilungen des Reichsamts für Landesaufnahme
Berlin. 1925–[1938]. 6 nos. a year. [762]
Although devoted primarily to German cartography there is often much cartographic information for other countries.

Behrmann, Walter. "Neuere deutsche länderkundliche Atlanten," *Zeitschrift für Erdkunde* [723], 4 (Feb., 1936), 97–103. [763]
A survey of 6 atlases of Germany or parts thereof, published between 1926 and 1936.

Deutscher Landwirtschaftsatlas. Bearbeitet im Statistischen Reichsamt. Berlin, Reimar Hobbing, 1934. 104 pl.; 34 pp. 37 x 45 cm. [764]
Crops, animal husbandry, forestry, rural economy, etc.

SWITZERLAND

Mudge-Winchell [1]: bibliographies, national and trade, a 412–413; bibliographies, statistics and economics, c 26, d 25; bibliographies, historical, a 360; dissertations, a 33; commercial handbook (1921), a 150; geographical names, a 331; periodicals, bibliographies of, a 24; statistical reference books, a 129, c 26.—See also [1171]. [765]

Bibliographie der schweizerischen naturwissenschaftlichen und geographischen Literatur. Berne, Schweizerische Landesbibliothek, 1927–. Annual [?]. [766]
Beginning with volume for 1940, published in 1942, *Geographische*

Bibliographie der Schweiz [768] was combined with this and "und geographischen" added in title (see Winkler [767], pp. 122–123, 153).

Winkler, Ernst. "Zur Bibliographie und Systematik der Schweizer Geographie: ein Rück- und Ausblick," *Mitteilungen der Geographisch-Ethnographischen Gesellschaft Zürich* [776], 41 (1941–42 & 1942–43), 122–158. [767]
Contains on pp. 150–158 a list of bibls. pertaining to Switzerland: (1) genl. bibls.; (2) professional bibls. (topical cl.); (3) regional bibls. (place cl.).

"Geographische Bibliographie der Schweiz," *Mitteilungen der Geographisch-Ethnographischen Gesellschaft in Zürich* [776], Vols. 20–40, 1919/20–1940/41. [768]
This annual bibl. of current literature covering 1919–1940, was compiled by August Aeppli for 1919–36 and by E. Winkler for 1937–40. See [766].

Hassinger, H. "Schweiz (1925–28)," *Geog. Jhrb.* [6], 43 (1928), 236–275. [769]
By topics, with a section on regions; 678 entries; ann.; auth. index.

Bibliographie der schweizerischen Landeskunde. . . . Herausgegeben von der Centralkommission für schweizerische Landeskunde. Berne, K. J. Wyss, 1892–[1927]. Also with title *Bibliographie nationale suisse.* [770]
A monumental series of bibls., for the most part published prior to the First World War, covering a wide variety of topics. The fascicules are cl. in 5 subseries dealing with (1) bibl. aids, societies, periodicals, etc. (1894–1896); (2) surveys and cartography (1892–1896); (3) travels and description, 1479–1900 (1899–1909); (4) natural hist. (1894–1927; the fascicule on climatology was published in 1927); (5) inhabitants (i.e. anthropology, economics, culture, etc., 1892–1920). Besterman [3], Vol. 2, pp. 356–358, lists the parts published.

———

Knapp, Charles, Maurice Borel, V. Attinger, and others. *Dictionnaire géographique de la Suisse.* Publié sous les auspices de la Société Neuchâteloise de Géographie, Neuchâtel, Attinger frères, 1902–1910. 6 vols. [771]
This magnificent geog. encyclopedia gives bibl. refs. in connection with the more important arts. There is also a German edition.

Periodicals

Le Globe

Société de Géographie de Genève. 1860–[1944]. This is published in 2 parts, *Bulletin*, semiannually, and *Mémoires*, annually. Indexes: for Vols. 1–50 (1860–1911); Vols. 51–60 (1912–1921) in *Bull.* for Nov., 1920–Apr., 1921; Vols. 61–70 (1922–1931) in *Mémoires*, Vol. 70 (1931).   [772]

Jahresbericht der Geographischen Gesellschaft von Bern

1878–. Normally biennial.   [773]

Mitteilungen der Ostschweizerischen Geographisch-Commerciellen Gesellschaft in St. Gallen

1878–[1940–41]. Annual.   [774]

Bulletin de la Société Neuchâteloise de Géographie

1885–. Annual.   [775]
Book reviews and monographs.

Mitteilungen der Geographisch-Ethnographischen Gesellschaft in Zürich

1917–1945. Annual. Succeeds *Jahresbericht*, 1899–1917. Succeeded by [778].   [776]

Der Schweizer Geograph: Le Géographe suisse: Zeitschrift des Vereins Schweizerischer Geographielehrer, Sowie der Geographischen Gesellschaften von Basel, Bern, St. Gallen und Zürich

Bern, 1923–[1944]. Bimonthly. Succeeded by [778].   [777]
Includes French as well as German contributions. Mainly devoted to geog. education.

Geographica helvetica: Schweizerische Zeitschrift für Länder- und Völkerkunde   [778]
Bern, Kummerly and Frey. 1946–. Quarterly. Continuation of [776, 777]. Edited by Geographisch-Ethnographische Gesellschaft Zürich with the coöperation of Geographische Gesellschaft Bern and Société de Géographie de Genève.

Mémoires de la Société Fribourgeoise des Sciences Naturelles: Géologie et Géographie

1900–1932. Occasionally.   [779]
Geog. monographs.

Maps and Atlases

Grob, Richard. "Geschichte der schweizerischen Kartographie," Part

I, in *Jahresbericht der Geographischen Gesellschaft von Bern,* 33 (1937/1939; published 1940), Pt. 1, 3–98. [780]
Covers per. from Ptolemy to the early 19th century. Many bibl. refs. in footnotes.

Graphisch-statistischer Atlas der Schweiz: Atlas graphique et statistique de la Suisse. Berne, Bureau de Statistique du Departement Fédéral de l'Intérieur, 1914. 65 pl.; 6 pp. 41 x 28 cm. [781]
Land use, population, education, agriculture, industry, commerce, transportation, finance, etc. (maps and graphs).

AUSTRIA

Mudge-Winchell [1]: bibliographies, historical, a 348, c 69; dissertations, b 12; statistical yearbook, a 124.—Official maps: Thiele [97], 1938, pp. 274-275.—See also [1169]. [782]

Lichtenecker, Norbert. "Österreich (1912–29)," *Geog. Jhrb.* [6], 45 (1930), 204–242. [783]
Topical and place cl.; *ca.* 710 entries; ann.; auth. index.

Periodicals and Series

Mitteilungen der (k.k.) Geographischen Gesellschaft in Wien
Vienna. 1857–[1944]. Monthly. Index: 1857–1907. [784]
Though somewhat less pretentious than other periodicals of great national geog. societies, this includes, none the less, arts. of high scholarly and scientific importance.

Geographischer Jahresbericht über (aus) Österreich
Vienna. 1894–[1940]. Nominally annual. Vol. 19, 1938, was published by Geographisches Institut an der Universität in Wien. [785]
Series of monographs on the geog. of Austria and of the former Austro-Hungarian Empire with summaries of the literature.

Jahrbuch für Landeskunde von Niederösterreich
Vienna, Verein für Landeskunde von Niederösterreich. 1902–[1938]. Succeeded *Blätter,* 1865–1901, of same society, which also published *Monatsblatt,* 1902–1932 (1928–1932 as *Unsere Heimat*). Indexes: 1865–1880; 1881–1890, 1891–1927. [786]
Kartographische und schulgeographische Zeitschrift
Vienna, Kartographische Anstalt von G. Freytag und Berndt. 1912–1922. 10 numbers a year. [787]

Mitteilungen des k̨. u. k̨. Militärgeographischen Institutes
Vienna. 1881–1913. Annual. Index: 1881–1904 in Vol. 24,
1904. [788]
Devoted to Austro-Hungarian government surveys.

Deutsche Rundschau für Geographie und Statistik̨
Vienna, A. Hartleben. 1878–1915. Monthly. [789]
Zeitschrift für wissenschaftliche Geographie
Lahr, later Weimar, later Vienna. 1880–1891. 8 vols. were published. [790]
This periodical contains material of scientific quality and in spirit was
the predecessor of the *Geographische Zeitschrift* [719].

Wiener geographische Studien. Vienna. 1933–[1939]. Ed. by Prof.
Dr. Hermann Leiter. [791]
Abhandlungen der (k̨. k̨.) Geographischen Gesellschaft in Wien.
Vienna. 1899–1922. Irregularly. Vol. 11 (1914–1920) published in
1920; Vol. 12 in 1922. [792]

CZECHOSLOVAKIA

Mudge-Winchell [1]: bibliographies, a 350, 395–396, b 55, c 77; encyclopedias, a 45, 50, b 13, c 16, d 15; periodicals, bibliographies, a 21;
statistical yearbooks, a 125.—See also [1168]. [793]

Mikula, Hermann. "Tschechoslowakei, 1912–27," *Geog. Jhrb.* [6],
43 (1928), 111–134. [794]
By topics and regions; 494 entries; ann.; auth. index.

Periodicals and Series

*Sborník̨ Československé Společnosti Zeměpisné: Journal of the
Czechoslovak̨ Geographical Society*
Prague. 1895–. 8 (formerly 10) numbers (in 4) a year. Before
1920 subtitle *Sborník̨ České Společnosti Zeměvědné.* [795]
Contains book reviews.

Vojenský Zeměpisný Ústav: Výroční Zpráva. [Military Geographical Institute: Annual Report.] 1918/20–[1932/34]. Prague, 1921–
1936. [796]
Primary source of information concerning maps of Czechoslovakia.
Résumés in French.

Zeměpisné Práce řídí Prof. Jiří Král: Travaux géographiques rédigés par Prof. Jiří Král. Bratislava. 1930–[1938]. [797]
1 monograph in German, 11 in Czech (with résumés in French except for 1 in German) dealing for the most part with Carpathian Czechoslovakia.

Arbeiten des Geographischen Institutes der Deutschen Universität in Prag. New series, 1921–[1934]. Irregularly. Earlier issues were reprints (see Introduction in No. 1). [798]

Travaux géographiques tchèques édités par le Prof. Dr. V. Švambera. Prague, Institut Géographique de l'Université Charles IV. 1901– [1932]. Previous to No. 11, 1923: Institut Géographique de l'Université Tchèque de Prague. [799]
Monographs on geog. of Czechoslovakia and by Czech geographers.

Atlas

Atlas Republiky Československé: Atlas de la République Tchécoslovaque. (L'Académie Tchèque.) Prague, Editions Orbis, 1935. 55 pl. 42 x 43 cm. Separate volume of text, by Václav Láska, published in French (44 pp.) and English (42 pp.) eds. [800]
An immense amount of information is provided by the superb colored plates (many with 8 maps apiece) of this atlas, covering physical and economic geog., demography, society, and culture. Ranks with *Atlas de France* [706] as one of the most comprehensive, detailed, and scholarly of all national atlases.

POLAND

Mudge-Winchell [1]: bibliographies, a 408, b 56, c 78, d 73; bibliographies, historical, a 359; encyclopedias, a 49, b 14; gazetteer, a 326; societies, a 37; statistical yearbooks, a 127–128. [801]

Plaetschke, Bruno. "Das geographische Schrifttum über Polen (1929–36)," *Geog. Jhrb.* [6], 51 (1936), 313–357. [802]
By topics, with a section on regions; 567 entries; ann.; auth. index.

Periodicals and Series

Ziemia [The Earth]

Warsaw, Polskie Towarzystwo Krajoznawcze [Polish Society for "Landeskunde"]. 1910–[1939]. Normally monthly. Index: 1910–1929. [803]

Semipopular illustrated periodical devoted to geog., culture, hist., etc., of Poland.

Czasopismo Geograficzne. . . . *Organe trimestriel de l'Association des Professeurs de Géographie*. Lwów. 1923-[1939]. Quarterly. Subtitle varies. [804] Long signed reviews.

Wiadomości Geograficzne: Bulletin trimestriel de géographie Cracow. 1923-[1939]. Subtitle, 1923–1938, was *Revue mensuelle de géographie*. Through 1936 published by the Cracow Section of the Polish Geographical Society. [805]

Wiadomości Służby Geograficznej: Kwartalnik Wojskowego Instytutu Geograficznego: Bulletin du Service Géographique: Revue trimestrielle de l'Institut Géographique Militaire à Varsovie Warsaw, 1927-[1939]. Quarterly. [806]

Przegląd Geograficzny: Revue polonaise de géographie Warsaw, Polskie Towarzystwo Geograficzne [Polish Geographical Society]. 1918–. Annual Index: 1918–1930. [807] Arts., notes, and book reviews. French résumés of the arts.

Polski Przegląd Kartograficzny: The Polish Cartographical Review (Quarterly); *La Revue cartographique polonaise (trimestrielle)* Lwów (Lemberg) and Warsaw. 1923–1934. Ed. by Professor Eugeniusz Romer. [808] Critical notes in Polish with brief résumés in French or English of maps and publs. on the cartography of Europe, Poland, and the adjacent countries.

Pamiętnik Fizyjograficzny [Physiographic Memoirs] Warsaw, Polskie Towarzystwo Krajoznawcze [Polish Society for "Landeskunde"]. 1881–1922. [809]

Prace Geograficzne Wydawane Przez E. Romera: Travaux géographiques publiés sous la direction de E. Romer. Lwów and Warsaw. 1918-[1938]. Irregularly. [810] Monographs dealing for the most part with the geog. of Poland.

Prace Instytutu Geograficznego Uniwersytetu Jagiellońskiego: Travaux de l'Institut Géographique de l'Université de Cracovie. Cracow. 1923-[1938]. Irregularly. [811] Monographs on the geog. of Poland.

Badania Geograficzne: Prace Instytutu Geograficznego Uniwersytetu Poznańskiego . . . : *Études géographiques: Travaux de l'Institut Géographique d'Université à Poznań.* . . . Poznań. 1926–[1937]. Irregularly.   [812]
In Polish with résumés in French or German.

Prace Wykonane w Zakładzie Geograficznym Uniwersytetu Warsawskiego: Travaux exécutés à l'Institut de Géographie de l'Université de Varsovie. Warsaw, 1922–[1937]. Irregularly.   [813]
Monographs.

Maps and Atlases

Joerg, W. L. G. "The Development of Polish Cartography Since the World War," *Geog. Rev.* [476], 23 (1933), 122–129.   [814]

Martonne, Ed. de. *Les Progrès de la carte en Pologne.* (Reprinted from *Bulletin de la Société de Topographie de France.* N.d. [c. 1936]. 16 pp.   [815]

—

Rzeczpospolita Polska, Atlas statystyczny: La République Polonaise, Atlas statistique. (Office Central de Statistique de la République Polonaise.) Warsaw, 1930. 42 pl.; 16 pp. 27 x 44 cm.   [816]
Text and legends on maps in Polish and French. Deals with Poland within the boundaries of 1918–1939. Population, agriculture, transportation, trade, labor, finance, education, etc. (maps and graphs).

Romer, Eugeniusz. *Geograficzno-statystyczny atlas Polski: Atlas de la Pologne (géographie et statistique).* Warsaw and Cracow, Gebethner & Wolff, 1916. 32 pl.; 32 pp. 35 x 32 cm.   [817]
Title, text, and legends on maps in Polish, French, and German. Covers not only the Poland of 1918–1939 but also a large surrounding area, especially to the east. Physical, economic, and historical geog., population, etc. See review by R. H. Lord in *Geog. Rev.* [476], 11 (1921), 308–309.

HUNGARY

Mudge-Winchell [1]: bibliographies, a 404; encyclopedias, a 48; statistical yearbook, a 126.—Official maps: Thiele [85], 1938, pp. 276–277.   [818]

Cholnoky, J. V. "Ungarn (1910–28)," *Geog. Jhrb.* [6], 43 (1928), pp. 181–192.   [819]
121 entries.

Bibliographia Hungariae. . . . *Verzeichnis der 1861–1921 erschienenen, Ungarn betreffenden Schriften in nichtungarischer Sprache,* constituting Vols. 1 and 2 of *Ungarische Bibliothek.* . . . Ungarische Institut an der Universität Berlin, Ser. 3. Vol. 1, *Historica,* Berlin, de Gruyter, 1923, 318 cols.; Vol. 2, *Geographica Politicooeconomica,* 1926, cols. 319–710.   [820]
Vol. 1 in Amer. Geog. Soc. Library; ref. to Vol. 2 from Besterman [3], Vol. 1, p. 481.

Periodicals

Földrajzi Közlemények: Geographical Review
    Budapest, Magyar Földrajzi Társaság: Hungarian Geographical Society. 1873–[1939]. 10 numbers a year. Quarterly in 1939. [821]
    Bibl. supplements covering 1936 and 1937 were published in 1938 and 1939 respectively.

Bulletin internationale de la Société Hongroise de Géographie
    Budapest. 1873–[1939]. (Publ. suspended 1920–1928.) Irregularly.   [822]
    Arts. in French, English, German, Italian, and Magyar.

A Földgömb: A Magyar Földrajzi Társaság Népszerü Folyóirata [The Terrestrial Globe: Popular Magazine of the Hungarian Geographical Society]
    Budapest. 1930–[1940]. Monthly.   [823]
Térképészeti Közlöny: Mitteilungen des Kgl. Ungarischen Kartographischen Institutes
    Budapest. 1930–[1937]. Also *Sonderheft* 1–6, 1930–1933. Biennially.   [824]

Iberian Peninsula and Italy

PORTUGAL

Mudge-Winchell [1]: bibliography, a 409; encyclopedias, a 49; statistical yearbook, a 128.   [825]

Lautensach, Hermann. *Portugal auf Grund eigener Reisen und der Literatur,* constituting *Petermanns Mitteilungen Ergänzungsheft* [718] *213,* 1932, and *230,* 1937.   [826]
    Bibl. in *Ergänzungsheft 230,* pp. 145–159 (topical and place cl.; 577 entries).

Lautensach, Hermann. "Portugal," *Geog. Jhrb.* [6], 45 (1930), 178–203. [827]
Sections on Portugal as a whole (topical cl.) and on regions (place cl.); 256 entries; ann.; auth. index.

Lautensach, Hermann. "Stand und Aufgaben der Landeskunde von Portugal," *Petermanns Mitteilungen: Ergänzungsheft* [718] 209 (1930), pp. 367–379. [828]
Bibl. refs. in the footnotes.

Anselmo, António. *Bibliografia das bibliografias portuguesas.* Lisbon, Biblioteca Nacional, 1923. 158 pp. (*Biblioteca do Bibliotecário e do Arquivista,* 3.) [829]
By topics, incl. (1) genl. bibls. and (2) special bibls., incl. geog., geology, cartography, hist., natural hist., travels; 1,030 entries; indexes to auths., subjs., archives, and libraries.

Periodical

Boletim da Sociedade de Geografia de Lisboa
Lisbon. 1876–. Normally monthly. [830]

SPAIN

Mudge-Winchell [1]: bibliographies, a 409–410, c 78; bibliographies, historical, a 360; encyclopedias, a 50, b 14, c 16, d 15; periodicals, bibliographies of, a 22, b 11; statistical yearbook, a 128. [831]

Panzer, Wolfgang. "Spanien (1915–30)," *Geog. Jhrb.* [6], 45 (1930), 133–177. [832]
Topical and place cl.; 839 entries; ann.; auth. index.

Periodicals

Boletín de la Real Sociedad Geográfica
Madrid. 1876–. Monthly; formerly quarterly. Not published 1937–1940. Indexes: 1876–1900; 1901–1910; 1911–1920; 1921–1930 in: J. M. Tarroja, *Repertorio de las Publicaciones y Tareas de la Real Sociedad Geográfica, 1921–1930.* [833]
Revista de Geografía Colonial y Mercantil
Madrid, Real Sociedad Geográfica. 1897–1924. Monthly. [834]
Includes geog. bibls. of some pretensions. Of especial utility for Spanish colonies in Africa.

ITALY

(See also [18, 1167], geographical bibliographies; [160], geographical studies.)

Mudge-Winchell [1]: bibliographies, a 405–406, b 56, c 78; encyclopedias, a 48–49, b 13–14, c 16; gazetteers, etc., a 325; government documents, a 11; libraries, a 421; periodicals, bibliographies of, a 22, b 11; periodicals, indexes, a 10–11, b 9; societies, a 37.   [835]

Migliorini, Elio, and others. *Bibliografia geografica della regione italiana*. Rome, R. Società Geografica Italiana, 1925–[1939]. Annual. The first 3 numbers (*fasciocoli*), entitled *Bibliografia geografica dell'Italia* and covering the years 1925–1927, were issued under the auspices of Comitato Geografico Nazionale Italiano and published in Florence by Istituto Geografico Militare. Subsequent numbers carry title as given in the heading above; those for 1928–1937 appeared each year in *Boll. R. Società Geografica Italiana* [841], 1929–1938; for 1938 and 1939 they were published separately by the R. Società Geografica Italiana.   [836]
Reviews comparable to those in *Geog. Jhrb.* [6]; running comment (topical cl.) with bibl. data in footnotes.

Almagià, Roberto. "Italien (1925–30)," *Geog. Jhrb.* [6], 46 (1931), 137–202.   [837]
Topical and regional cl.; 908 entries; ann.; auth. index.

Magistris, L. F. de. *Bibliografia geografica della regione italiana: saggio per l'anno 1899*. Rome, Società Geografica Italiana, 1901.   [838]
Continuation of regional bibls. previously published in the *Bollettino della Società Geografica Italiana* [841]. It was continued for the period through the year 1903 in *Rivista geografica italiana* [843].
Running commentary on progress of regional investigations in Italy with bibl. data in footnotes.

Hellman, Florence S. *Sicily and Sardinia: A Bibliographical List.* (The Library of Congress, Division of Bibliography.) Washington, 1942, 38 mimeo. pp.   [839]
Sections on Sicily and Sardinia, each cl. by topics, incl. bibls., geog., geol., hist., people, economics, etc.; 455 entries.

Periodicals and Series

Atti degli Congressi Geografichi Italiani
    1892–1937. Index, 1892–1927.   [840]
    The following list shows the places and dates of the Congresses and,
    in parentheses, the dates of the publ. of the *Atti,* the place of publ.
    regularly being the same as that of the Congress:
        First, Genoa, 1892 (1893–94); Second, Rome, 1895 (1896); Third,
    Florence, 1898 (1899); Fourth, Milan, 1901 (1902); Fifth, Naples, 1904
    (1905); Sixth, Venice, 1907 (1908); Seventh, Palermo, 1910 (1911);
    Eighth, Florence, 1921 (1922); Ninth, Genoa, 1924 (1925–27); Tenth,
    Milan, 1927 (1927); Eleventh, Naples, 1930 (1930); Twelfth, Sardinia,
    1934 (1935); Thirteenth, Friuli, 1937 (1938).

Bollettino della Reale Società Geografica Italiana
    Rome. 1868–. Monthly. Through 1912 the title was *Boll. della
    Società Geog. Ital.;* 1913–June, 1916, *Boll. della Reale Società Geo-
    grafica;* since June, 1916, as at present. Indexes: 1868–75; 1876–87;
    1888–99; 1912–23.   [841]
    Contains book reviews and lists of accessions to library.

Memorie della Reale Società Geografica Italiana
    Rome. 1878–. Irregularly.   [842]

*Rivista geografica italiana: Organo della "Società di Studi Geografici"
residente in Firenze*
    Florence. 1894–[1944]. Monthly. Subtitle varies. Index: 1894–
    1903.   [843]
    The leading critical Italian geog. period. From Jan., 1914, through 1919
    book reviews and a topically arranged bibl. of geog. publs., by Roberto
    Almagià, were published bimonthly in a Supplement to the *Rivista* en-
    titled *Rassegna della letteratura geografica.* From 1920 to 1938 inclusive
    this *Rassegna* was included in the *Rivista.*

L'Universo
    Florence, Istituto Geografico Militare. 1920–[1941]. Monthly,
    formerly bimonthly. Supersedes *Annuario dell'Istituto Geografico
    Militare,* 1913–1915.   [844]
    Specializes in mathematical geog., surveying, cartography, photogram-
    metry. Incl. reviews.

*Geopolitica: Rassegna mensile di geografia politica, economica, so-
ciale, coloniale*
    Milan, Sperling & Kupfer. 1939–[1940]. Monthly.   [845]

Pubblicazioni dell'Istituto di Geografia della R. Università di Trieste
Trieste. 1942-[1943].   [845a]

Rivista di geografia
Milan and Florence, D. Bellazzi. 1916–1928, New Ser. 1932–1933.
Irregularly. Prior to 1929, *Rivista di geografia didattica;* Apr.–Dec.,
1932, *Rivista di geografia e cultura geografica.* Supplement to the
Rivista geografica italiana [843] devoted to geog. education. [846]

La Geografia
Novara, Istituto Geografico De Agostini. 1912–1930. 10 numbers
to the vol., 1914–1917; 6, 1918–1930.   [847]
Brief bibl. notes.

Atti della Società Ligustica di Scienze Naturali e Geografiche
Genoa. 1890–1920. Irregularly.   [848]
Occasional geog. articles.

Memorie geografiche
Florence. 1907–1919. Published as a supplement to the *Rivista geografica italiana* [843].   [849]
Monographs dealing, in large measure, with the physiography, hydrography, orometry, climatology, human geog., and hist. of cartography
of Italy and the hydrography of the Mediterranean basin.

Atlases

Giotto Dainelli. *Atlante fisico-economico d'Italia.* Esecuzione tecnica
del Laboratorio Cartografico della C.T.I., diretto da Piero Corbellini. Milan, Consociazione Turistica Italiana, 1940. 82 pl.; 17 pp.
31 x 49 cm. Accompanied by volume of text (*Note illustrative*) by
Aldo Sestini, 1940. 147 pp.   [850]
Physical and economic geog., demography, culture, and administration
of Italy. A plate for Italians abroad. See *Geog. Rev.* [476], 30 (1940),
340–342.

Atlante statistico italiano. (Istituto Centrale di Statistica del Regno
d'Italia.) Bergamo, Istit. Italiano d'Arti Grafiche. 2 vols. (*Parte*):
(1) *Natalità, mortalità, densità della popolazione, 1929;* (2) *Nuzialità, aumento della popolazione dal 1911 al 1921, frazionamento
della proprietà terriera, reddito, 1933.* 265 pl. 61 x 52 cm.   [851]
11 colored cartograms of Italy on the scale of 1:500,000, each subdivided into 17 sections (plates); one cartogram, 1:1,000,000, sub-

divided into 5 sections; in Part 2 a key map showing relief, drainage, and town names faces each cartogram.

Eastern and Southeastern Europe

The Balkan Peninsula: Mudge-Winchell [1]: bibliographies, a 348; statistical yearbook, a 124.—Greece: Mudge-Winchell [1]: bibliographies, a 403, c 78; encyclopedias, a 48, b 13.   [852]

Kerner, R. J. *Slavic Europe: A Selected Bibliography in the Western European Languages, Comprising History, Languages, and Literatures.* Cambridge, Mass., Harvard University Press, 1918. 426 pp.   [853]
Valuable for the geographer as well as for the historian. Sections on the Slavs, the Russians, the Poles, the Slavs in the German Empire, the Bohemians and the Slovaks, and the Southern Slavs, each subcl. by topics, incl. lesser ethnic groups, bibls., hist., geog., cartography, language, literature, etc.; 4,521 entries; auth. index.

Osteuropäische Bibliographie. Osteuropa Institut in Breslau, 1920–1923. Annual. Vols. 1 and 2, Leipzig and Berlin, Teubner, 1921, 1923; Vols. 3 and 4, Breslau, Priebatsch, 1926, 1928.   [854]
By countries, subcl. by topics incl. geog., "state and society," economics, law, religon, hist., literature; auth. index. See comment in Coulter and Gerstenfeld [137], p. 78.

————

Conover, Helen F. *The Balkans: I, General: A Selected List of References.* (The Library of Congress, Division of Bibliography.) Washington, 1943. 73 mimeo. pp.   [855]
This and the other four bibls. in this series (see [860, 867, 873, 879] are restricted in the main to works published since the Treaty of Versailles, and, with a few exceptions, to writings in the languages of western Europe available in American libraries. By topics, incl. bibls., genl., the land, political hist., economics, people; auth. index; genl. subj. index to the five Balkan lists.

Savadjian, Léon. *Bibliographie balkanique.* Paris, Société Générale d'Imprimerie et d'Édition, 1920–1938. 8 vols.: for 1920–1930 (published by *Revue des Balkans*), 1931–1932, and annually, 1933–1938.   [856]
Deals with works published in French, Italian, German, and Englis¹
By countries; auth. index.

Weiss, Jakob. "Bericht über die Länder- und Völkerkunde Sudost-Europas im Rahmen des antiken Geographie (1911–27)," *Geog. Jhrb.* [6], 43 (1928), 135–179.   [857]
Covers Greece, Macedonia, and Thrace. Primarily, though not exclusively, historical and archeological. Place cl., subcl. by topics; 225 entries; ann.; auth. index.

Cvijić, Jovan. *Pregled geografske literature o Balkanskom poluostrvu: Bibliographie géographique de la péninsule balkanique.* Belgrade, 1894–1908. 5 vols. In Serbian.   [858]
Covers the per. 1891–1905. Topical cl.; ann.; geog. and auth. indexes. For a commendatory notice see *Ann. de Géog. Bibl.* [7], 18 (1908), 164–165.

YUGOSLAVIA

Mudge-Winchell [1]: bibliographies, a 406; encyclopedia, a 50; commercial handbook (1928), a 150; national handbook, b 22.   [859]

Conover, Helen F. *The Balkans, V. Yugoslavia: A Selected List of References.* (The Library of Congress, Division of Bibliography.) Washington, 1943. 63 mimeo. pp.   [860]
By topics, incl. bibls., genl., the land, hist. and politics, economics, people; 605 entries; auth. index. See also [855].

Vujević, P. "Südslawien 1913–28," *Geog. Jhrb.* [6], 45 (1929), 252–288.   [861]
Topical cl., incl. a short section on regions; 741 entries; ann.; auth. index.

Periodicals

Geografski Vestnik: Časopis za Geografijo in Sorodne Vede: Bulletin de la Société de Géographie de Ljubljana
1925–. Quarterly.   [862]
Arts. in Croat (or German) with French or German résumés.

Hrvatski Geografski Glasnik: Croatian Geographical Review
Zagreb, 1929–[1939]. Ed. by Prof. Artur Gavazzi.   [863]
Arts. in Croat (or some occidental language) with French or German résumés. Book reviews.

Glasnik Geografskog Drushtva: Bulletin de la Société de Géographie de Beograde
1912–[1938]. Annual. Not published, 1915–1920.   [864]

In Serbian with occasional résumés in French. A scholarly and scientific publ. Bibls.

Mémoires de la Société de Géographie de Beograd
1933-[1936]. Annual. [865]

RUMANIA

Mudge-Winchell [1]: bibliographies, a 409, b 56; commercial handbook (1924), a 150; statistical yearbooks, a 128, c 26. [866]

Conover, Helen F. *The Balkans, IV. Rumania: A Selected List of References*. (The Library of Congress, Division of Bibliography.) Washington, 1943. 70 mimeo. pp. [867]
By topics, incl. bibls.; official publs., genl., the land, hist., economics, people; 661 entries; auth. index. See also [855].

Wachner, Heinrich. "Rumänien (1929-37)," *Geog. Jhrb.* [6], 53, Pt. 2 (1938), 631-686. [868]
Topical and place cl.; 733 entries; ann.; auth. index.

Periodicals and Series

Buletinul Societății Regale Române de Geografie
Bucharest. 1876-[1938]. Annual since 1919; previously semi-annual. [869]
Extensive book reviews and bibls. French summaries of many of the arts.

Lucările Institutului de Geografie al Universității din Cluj: Travaux de l'Institut de Géographie de l'Université de Cluj. 1922-[1938]. Irregularly. [870]

Atlas

L'Agriculture en Roumanie: Atlas statistique. Publié par le Ministère de l'Agriculture et des Domaines. Bucharest, Imprimeria Națională, 1938. 99 pp. (maps on right-hand, statistics, graphs, and text on left-hand, pages). 42 x 32 cm. [871]
Crops, animal husbandry, and rural economy; also climate, geology, and vegetation.

BULGARIA

Note on some recent studies, regional and agrarian, by Raymond Crist, in *Geog. Rev.* [476], 30 (1940), 495–497.   [872]

Conover, Helen F. *The Balkans, III. Bulgaria: A Selected List of References*. (The Library of Congress, Division of Bibliography.) Washington, 1943. 34 mimeo. pp.   [873]
By topics, incl. bibls., official publs., genl., the land, hist., economics, people; 322 entries; auth. index. See also [855].

Wilhelmy, Herbert. "Bulgarien (1910–32)," *Geog. Jhrb.* [6], 48 (1933), 51–100.   [874]
Topical and place cl.; 783 entries; ann.; auth. index.

Periodicals

Izvestiia na B"lgarskoto Geografsko Druzhestvo: Bulletin de la Société Bulgare de Géographie
Sofia. 1933–[1939]. Annual. Title also in German (*Mitteilungen*).   [875]
Arts. in Bulgarian. Résumés in French or German. Contains book reviews.

Godishnik na D"rzhavniia Geografski Institut pri Ministerstvoto na Voïnata [Yearbook of the State Geographical Institute of the Ministry of War]
Sofia. 1922–[1932].   [876]

Atlas

The Bulgarians in Their Historical, Ethnographical and Political Frontiers. Preface by D. Rizoff. Berlin, Greve, 1917. 40 pl., with accompanying text in German, English, French and Bulgarian. 30 x 30 cm.   [877]
Earlier historical, political, and ethnographic maps, reproduced for purposes of Bulgarian propaganda during World War I.

ALBANIA

Mudge-Winchell [1]: bibliographies, national, a 392.   [878]

Conover, Helen F. *The Balkans, II. Albania: A Selected List of References*. (The Library of Congress, Division of Bibliography.) Washington, 1943. 24 mimeo. pp.   [879]

By topics, incl. bibls., official publs., genl., the land, hist. and politics, economics, people; 239 entries; auth. index. See also [855].

Istituto di Geografia della R. Università degli Studi "Benito Mussolini." *Saggio di bibliografia geografica d'Albania.* (Separate from *Annali della Facoltà di Economia e Commercio della R. Università di Bari,* N.S. Vol. 2.) Bari, A. Cressati, 1939. 88 pp.   [880]
By chron. pers. (–1912, 1913–1924, 1925–1938), each subcl. by topics; ann.

UNION OF SOCIALIST SOVIET REPUBLICS

U.S.S.R. in General and in Europe

(See also [161, 164], geographical studies.)

Mudge-Winchell [1]: encyclopedias, a 49–50, b 14, c 16; periodicals, bibliographies of, a 22; periodicals, indexes, a 11; statistical reference books, a 128, b 22.—Soils: Giesecke [230], 1939, pp. 260–266.—Economic geography: Lütgens [284], 1936, pp. 77–89. [881]

C. D. Harris, of the Department of Geography, University of Chicago, has recently prepared a mimeographed list of works in the University of Chicago libraries dealing with the geography of the Soviet Union. Cl. by topics and regions, it includes references to works in Russian as well as to books and periodical articles in other languages.   [882]

Conover, Helen F. *Soviet Russia: A Selected List of Recent References.* (The Library of Congress, Division of Bibliography.) Washington, 1943. 85 mimeo. pp.   [883]
Refs. to works in English for per. since 1937; by topics, incl. bibls., periods., description and travel, the land (subcl. by regions), hist., etc.; 811 entries; auth. index.

Grierson, Philip. *Books on Soviet Russia 1917–1942: A Bibliography and a Guide to Reading.* London, Methuen, 1943. 369 pp.   [884]
Covers books and pamphlets on postrevolutionary Russia that have been published in Great Britain, with a few refs. to works published elsewhere.
  By topics, incl. bibls., guide books, periods., the revolution and civil war, the Soviet state, economic life, culture and social life (with a sec-

tion "Soviet Geography and Exploration," place cl.), fiction (translations); ann.; auth. index. with a few subj. hdgs.

Schultz, Arved. "Europäisches Russland (1929–36)," *Geog. Jhrb.*
[6], 52 (1937), 75–248.   [885]
One of the most comprehensive and detailed of the regional surveys
in recent issues of the *Geog. Jhrb.* By topics (incl. progress reports on
geog. research, periods., maps, physical and biological geog., population, economic geog.) and by regions, subcl. by topics; 2,838 entries;
ann.; auth. index.

Anger, Helmut. *Die wichtigste geographische Literatur über das
Russische Reich seit dem Jahre 1914 in russischer Sprache.* Hamburg, 1926. 64 pp. (*Veröffentlichungen des Geographischen Instituts der Albertus-Universität zu Königsberg,* No. 6.)   [886]
Though a list of Russian works, the titles are given only as translated
into German. By topics, incl. bibls., periods., maps, and various aspects
of geog.; a few anns.

Berg, L. S. *Ocherk Ístoríí Russkoǐ geografícheskoǐ nauḳí (vplot'
do 1923 goda)* [Outline of the History of Russian Geographical
Science (to the year 1923)]. Leningrad, 1929. 154 pp. (*Akademiia
Nauk S.S.S.R.: Trudy Komíssíí po Ístoríí Znaníí* [Academy of
Science U.S.S.R.: Transactions of the Commission on the History
of Science], 4.)   [887]
While dealing primarily with per. prior to 1923, includes additional
data to December, 1926. Although not a bibl., it contains a wealth of
bibl. refs. Chapters on geog. methodology, institutions and publs. (including a section on bibls., pp. 18–20), cartography, explorations by
land, maritime explorations, historical geog.

Rudnyckyj, Stephan. "Einige Bemerkungen über die landeskundliche Literatur Russlands," *Mitteilungen k.k. Geographischen Gesellschaft in Wien* [784], 59 (1916), 615–632.   [888]
A brief but critical account of studies in Russian geog., geology, hist.,
ethnography, economics, etc.

Semenov", V. P., and others, eds. *Rossiia.* St. Petersburg, 1899–
1914.   [889]
According to the original plan this was to be a great coöperative descriptive work in 20 vols. on the entire Russian Empire, to be fully illustrated
with photographs, maps, etc. By 1914 vols. had appeared on the following regions: Vol. 1, Moscow industrial district and the upper Volga

basin; 3, The Ozernaya region; 6, Regions on both sides of the middle and lower Volga; 2, Central Russia and the Black Earth region; 5, Ural regions; 7, Little Russia; 9, The upper Dnieper and White Russia; 14, New Russia and Crimea; 16, Western Siberia; 18, The Kirghiz countries; 19, Turkestan. This work has been criticized rather severely; on the other hand, the bibl. sections (topical cl.) in each volume have been praised as the most valuable part. (See *Mitt. k.k Geogr. Gesell. in Wien* [784], 59 (1916), 616–617.)

Zelenīn", D. K. *Bibliograficheskiĭ ukazatel' Russkoĭ étnograficheskoĭ līteratury o vnieshnem" bytie narodov" Rossiī 1700–1910 g.g.* [Bibliographical Index of Russian Ethnographical Literature Relating to the External Mode of Life of the Peoples of Russia, 1700–1910], constituting *Zapīskī Īmperatorskago Russkago Geograficheskago Obshschestva po Otdieleniiu Étnografii* [900], 40, Issue 1. St. Petersburg, 1913. 772 pp.   [890]
This book serves as an introduction to the geog. as well as ethnographical literature of Russia of the per. prior to 1911. By topics, incl. genl. works, works on dwellings, clothing, music, art, and economic life, each minutely subcl. by topics and regions; list of ethnographical and geog. bibls. of Russia, pp. 3–9; 8,847 entries; place, ethnographic, map, and auth. indexes.

"Geograficheskaia līteratura: po dannym" bīblioteki Īmperatorskago Russkago Geograficheskago Obshchestva" [Geographical Literature: on the Basis of Data in the Library of the Imperial Russian Geographical Society], *Izviestiia I. Russkago Geograficheskago Obshchestva* [895], 41 (1905), 806–822; 42, No. 1 (1906), 333–352.   [891]

"Geograficheskaia līteratura: Rossiī Evropeĭskoĭ ī Aziatskoĭ ī prīlezhashchīkh" stran" [Geographical Literature: European and Asiatic Russia and Contiguous Countries], *Izviestiia Īmperatorskago Russkago Geograficheskago Obshchestva* [895], Vols. 27–35 (1891–1899).   [892]

Mezhov", V. Ī. *Līteratura russkoĭ geografii étnografii i statistīkī* [The Literature of Russian Geography, Ethnography, and Statistics]. 9 vols., covering 1859–1880, some of which were published in *Viestnik"* and *Īzviestiia Ī. R. Geograficheskago Obshchestva* [895], 1860–1883, and others separately.   [893]
Fundamental for per. covered. Topical cl.

Periodicals

Russian geographical periodicals and periodicals dealing with related subjects are listed in Anger [886], Schulz [885], and Leimbach [915]. For European U.S.S.R. Schulz, 1936, lists 311 titles under the following headings: general, 15; mathematical geography and cartography, 3; geology, mineralogy, soils, 38; geophysics, meteorology, climatology, 34; oceanography, hydrography, 16; plant geography, 14; animal geography, 9; anthropology, ethnology, 17; economics, statistics, 21; various other subjects, 24; regional, 136.

According to Schulz the only current (1936) general geographical periodicals of really scientific caliber are *Zemlevedenīe* [897] and *Izvestiia V. G. O.* [895].

The principal repository of geographical information on Russia in the collections of the American Geographical Society consists of the periodicals of the former Imperial Russian Geographical Society and its continuation (see [894, 895, 900, 901]; for changes in name see note on [895]), of many special publications of that Society, and of the periodicals of its various regional sections and subsections.

Between 1850 and 1917 about a dozen regional sections and subsections of the Imperial Russian Geographical Society were established in various parts of the Empire. All of these published one or more periodicals, some of which were continued to as late as 1930. The following partial list, which covers the outstanding series only, is based upon the check list of holdings in the library of the American Geographical Society, supplemented by data gleaned from Leimbach [915] and from the Union List of Serials (see also, for additional data, Kerner [963], Vol. 2, pp. 293–294). The list is arranged by institutions in the chronological order of the first number of the first periodical issued by each. Abbreviations: *Iz.*: *Izvestiia* [*Bulletin*]; *Zap.*: *Zapiski* [*Memoirs*]; (A): represented in the library of the American Geographical Society. See also the first edition of *Aids to Geographical Research* for titles of a selection of these periodicals as transliterated in full.

Caucasian Section, Tiflis: *Zap.*, 1852–1919 (A); *Iz.*, 1872–1917.

East Siberian (formerly Siberian) Section, Irkutsk: *Zap.*, 1856–1914; *Iz.*, 1870–1929 (A); *Biulleten'*, 1923–c1930.

Orenburg Section: *Zap.,* 1870–1881; *Iz.,* 1893–1916; *Trudy* [Transactions], 1928.

West Siberian Section, Omsk: *Zap.,* 1879–1927; *Iz.,* 1913–1930.

Amur Territory Section, Vladivostok Subsection (also has been known as Society for the Investigation of the Amur Region: Branch of the Imperial R. G. S., and (1922) as South Ussurī Section of the Russian G. S.): *Zap.,* 1888–1922, 1928–1929; *Iz.,* 1922.

Amur Territory Section, Khabarovsk: *Zap.,* 1894–1914(?).

Amur Territory Section, Troītskosavsk Subsection, Kiakhta: *Protokoly* [Records], 1894–1909; *Trudy,* 1898–1922.

Turkestan Section, Tashkent: *Iz.,* 1899–1929.

Amur Territory Section, Transbaikal Subsection, Chīta: *Zap.,* 1895–1928.

Krasnoiarsk Section: *Iz.,* 1901–1929; *Zap.,* 1902–1906.

Iakutsk (Yakutsk) Section: *Iz.,* 1915–1928.   [893a]

———

Ukazatel' k" izdaniiam" Imperatorskago Russkago Geograficheskago Obshchestva i ego otdielov" s" 1846 po 1875 god" [Guide to the Publications of the Imperial Russian Geographical Society and Its Branches for the Years 1846 to 1875]
   St. Petersburg, 1886. Similar guides covering 1876–1885, 1886–1895, 1896–1905 were published in 1887, 1896, and 1910 respectively.   [894]
Tables of contents with auth. and place indexes.

———

Izvestiia Vsesoiuznogo Geograficheskogo Obshchestva [Bulletin of the All-Union Geographical Society]: *Izvestia de la Société de Géographie de l'URSS*
   Moscow, Leningrad. 1865–. Recently 6 numbers a year (10 in 1939). Title: *Iz. Imperatorskago Russkago G. O.,* 1865–1916 (Indexes: see [894]; *Izv. Russkogo G. O.,* 1917–1924; *Izv. Gosudarstvennogo* [State] *Russkogo G. O.,* 1924–1930; *Izv. Gosudarstvennogo G. O.,* 1931–1939; as above, 1940.   [895]
Geog. monographs, bibls., book reviews. See comprehensive résumés of contents of recent issues in *Bibl. géographique internationale* [7].

Geografiia v Shkole [Geography in the School]
Moscow. 1934–[1939]. Irregularly.   [896]
For teachers.

Zemlevedenīe: GeograficheskīiZhurnal īm. D. N. Anuchīna [Geography: Geographical Journal named for D. N. Anuchin]
Moscow, Leningrad. 1894–[1938]. Normally quarterly. Ed.
(1938): A. A. Borzov. Subtitle varies.   [897]
Monographs, book reviews, bibl. notes. Russian counterpart of *Annales de géographie* [694] and *Geographische Zeitschrift* [719].

Īzvestīia Geograficheskogo Īnstītuta: Bulletin de l'Institut Géographique de Pétrograd
1919–[1926].   [898]

Geograficheskīi Vestnīk [Geographical Messenger] Petrograd, [Geographical Institute], 1922–[1925].   [899]
Monographs, geog. notes, books reviews, bibls.

Zapīskī Īmperatorskago Russkago Geograficheskago Obshchestva [Memoirs of the Imperial Russian Geographical Society]
St. Petersburg. 1861–1916. Irregularly. This series succeeded an earlier series of the same title, 1846–1859 (1860?), and a series entitled *Viestnīk" Īmperatorskago Russkago Geograficheskago Obshchestva* [*Messenger of the Imperial Russian Geographical Society*] published twice a year, 1851–1860. After 1866 the *Zapīskī* were published in 3 sections devoted respectively to geog., to statistics, and to ethnography (*po Obshcheī Geografīi; po Otdielenīiu Statīstīkī; po Otdielenīiu Ētnografīi*). Indexes: see [894].   [900]
Includes bibls., monographs, and reviews.

Ezhegodnīk" Īmperatorskago Russkago Geograficheskago Obshchestva [Yearbook of the Imperial Russian Geographical Society]
St. Petersburg. 1890–1899. 8 numbers were published. Indexes: see [894].   [901]
A Russian equivalent of the *Geographisches Jahrbuch* (*Geog. Jhrb* [6], 29 (1906), 150).

Cartography. (See also [121].)

Salīshchev, K. A. *Osnovy Kartovedenīia: Īstorīcheskaia Chast'* [Fundamentals of Cartography: Historical Part]. Moscow, 1943. 238 pp.   [902]

A history of cartography of especial value for the light shed on recent development of cartography in the U.S.S.R. (See *Geog. Rev.* [476], 35 (1945), 510.)

BALTIC STATES

Giere, Werner. "Die ostbaltischen Staaten: Litauen, Lettland, Estland (1928–36)," *Geog. Jhrb.* [6], 51 (1936), 358–418.   [903]
By topics and by countries subcl. by topics; 767 entries; ann.; auth. index.

Prinzhorn, F. *Memelgebiet und baltische Staaten: Eine Bibliographie mit besonderer Berüchsichtigung von Politik und Wirtschaft 1935/ 36 mit Nachträgen aus den Jahren 1931–34,* Vol. 1, Nos. 1–3. Danzig, 1936–1937. 63 pp.   [904]
By regions subcl. by topics, incl. land, population, hist., law, culture, economics, etc.

ESTONIA

Mudge-Winchell [1]: bibliographies, a 398; periodicals, bibliography of, a 21; statistical handbook, a 125.   [905]

Series

Publicationes Instituti Universitatis Dorpatensis [Tartuensis] Geographici. Dorpat [Tartu]. 1925–[1940]. Irregularly.   [906]
Most of these 24 monographs are in German (2 in English) and deal with the Baltic region and Estonia, but some are wider in scope.

Tartu Ülikooli Majandusgeograafia Seminari Üllitised: Publicationes Seminarii Universitatis Tartuensis Oeconomico-Geographici. Tartu and Talinn. 1931–[1939].   [907]
Monographs in Estonian, Swedish, German, and English dealing primarily with the economic geog. of Estonia. French, German, and English résumés.

Atlas

Schmidt, V. *Eesti: Statistiline album: Album statistique.* Bureau Central de Statistique de l'Estonie. [Talinn.] Vol. 1, 1925, 36 pl.; Vol. 2 [1927], 24 pl. 27 x 37 cm.   [908]
In Estonian and French. Physical and economic geog., demography, etc. (maps and graphs). See also: *Eesti pollumajandus statistiline album: Agriculture en Estonie.* Tallinn, Bureau Central de Statistique de l'Estonie, 1928. 112 pp. of maps, graphs, and text.

LATVIA

Mudge-Winchell [1]: bibliography, a 406; encyclopedias, a 49, b 14, c 16.   [909]

Periodical

Ģeōgrafiski Raksti: *Folia Geographica (Latvijas Ģeōgrafijas Biedrība: Societas Geographica Latviensis)*
Riga. 1928–[1938]. Irregularly.   [910]
French, German, or English résumés are given of some but not all of the arts. Record and review sections.

Maps

Krikščiunas, A. "Aus der Geschichte der geodätischen und kartographischen Arbeit in Litauen," in *Comptes rendus de la dixième session de la Commission Géodésique Baltique,* 1938, Helsinki, 1938, pp. 87–93.   [911]

LITHUANIA

Prinzhorn, Fritz, and Manfred Hellmann. "Schrifttum über Litauen (1928–38)," *Deutsches Archiv für Landes- und Volksforschung,* 3 (1939), 217–252.   [912]
By topics, incl. bibls., land, people, hist., politics, economics; ann.

Baltramaïtis", S. *Sbornik" bibliograficheskikh" materialov" dlia geografii, istorii, istorii prava, statistiki i ètnografii Litvy* [Collection of Bibliographical Materials for the Geography, History, Legal History, Statistics, and Ethnography of Lithuania]. St. Petersburg, Imperial Russian Geographical Society, 1904. 616 pp.   [913]
By topics, incl. geog., hydrography, geology, orography, travels, atlases, maps, etc.; 8,513 entries; auth. and subj. indexes.

U.S.S.R. in Asia

Bibliographies on Russia in Asia are listed in Kerner [963], 1939, Vol. 2, pp. 276–278, 292–293 (see also subject index under "Bibliography").   [914]

Leimbach, Werner. "Nordasien, Westturkistan und Innerasien (1926–37)," *Geog. Jhrb.* [6], 53, Pt. 2 (1938), 437–565; 54, Pt. 1 (1939), 303–352; 54, Pt. 2 (1939), 555–596.   [915]
2 parts: (1) Northern Asia (i.e. Siberia), in Vol. 53; (2) Western

Turkestan, in Vol. 54; each cl. by topics (incl. bibls., periods., physical, biological, and human geog.), subcl. by places; 1,728 entries; ann.; auth. index.

Mezhov"; V. Ī. *Sibīrskaia bibliografiia: Bibliographia Sibirica: Bibliographie des livres et articles de journaux russes et étrangers concernant la Sibérie.* St. Petersburg, 1891–1892. 3 vols. 1,446 pp. [916] Covers per. to 1890; 25,250 entries. Additions by A. A. Īvanovskiĭ in *Ėtnograficheskoe Obozrienie* [Ethnographic Review], 1892, Pt. 1, and *Bibliograficheskiia Zapīskī* [Bibliographic Records], 1892, Pt. 1. See also [927].

Aziatskaia Rossiia [Asiatic Russia]. Īzdanie Pereselencheskago Upravleniia Glavnago Upravleniia Zemleustroĭstva ī Zemledieliia [Publication of Dept. of Emigration of the Chief Administration of Land Distribution and Agriculture]. Vol. 3, St. Petersburg, 1914. [917] Bibl. on pp. lxxi–cxli: by topics, incl. agriculture, emigration, natural hist., genl. geog., cartography.

Mezhov", V. Ī. *Turkestanskiĭ sbornīk" sochīneniĭ ī stateĭ otnosiashchīkhsia do Srednei Aziī voobshche ī Turkestanskago Kraia v" osobennostī: Recueil du Turkestan, comprenant des ouvrages et des articles sur l'Asie central en général et le* [sic] *province du Turkestan en particulier.* St. Petersburg, 1878–1888. 3 vols. 506 pp. [918] 4,706 entries; auth., subj., and place indexes. Vol. 2 has French preface.

Atlas

Atlas" Aziatskoĭ Rossiī [Atlas of Asiatic Russia]. Īzdanie Pereselencheskago Upravleniia Glavnago Upravleniia Zemleustroĭstva ī Zemledieliia [Publication of the Dept. of Emigration of the Chief Administration of Land Distribution and Agriculture]. St. Petersburg, 1914. 71 pl.; 27 pp. (introduction and index). 41 x 53 cm. [919] Physical, historical, and economic geog., ethnography, etc., with especial emphasis on land settlement.

THE ORIENT, ISLAM, AND THE MEDITERRANEAN REGION

THE ORIENT

Orientalische Bibliographie. Berlin, Reuther und Reichard. 1887–
1911, 1926. Annual.   [920]
Covers all aspects of Oriental studies, including geog. Sections on genl.
works and areas subcl. by topics; auth. indexes. (See Mudge-Winchell
[1], a 281.)

Katalog der Bibliothek der Deutschen Morgenländischen Gesell-
schaft. 2d ed. Leipzig, 1900. Vol. 1, *Drucke.* 744 pp.   [921]
Although devoted in the main to Oriental philology, includes (pp. 660–
712) a list of books and reprints on geog. and ethnography (place cl.).

ISLAM

(See also [156], Turkish geography; [183], early cartography.)

Houtsma, M. T., and others, ed. *The Encyclopaedia of Islam: A Dic-*
tionary of the Geography, Ethnography and Biography of the
Muhammadan Peoples. Leiden (Brill), London (Luzac), 1913–
[1938]. 4 vols. with 5 Supplements.   [922]
Bibl. notes are appended to the geog. arts.

Pfannmüller, Gustav. *Handbuch der Islam-Literatur.* Berlin and
Leipzig, De Gruyter, 1923. 444 pp.   [923]
Lists of books followed by running descriptions. The aim is to include
only titles of works of lasting value. Main sections (each subcl. by topics)
devoted to bibls. of Islam in genl., Islamic lands and peoples, political
and cultural hist., religion, philosophy, art, and literature; auth. index.

Gabrieli, Giuseppe. *Manuale di bibliografia musulmana.* Part I,
Bibliografia generale. Roma, Tipografia dell'Unione Editrice, 1916.
491 pp. (*Manuali coloniali pubblicati a cura del Ministero delle*
colonie.)   [924]
By topics, incl. bibls., periods., Orientalism, education, etc.; auth. index.

THE MEDITERRANEAN REGION

East, Gordon. "The Mediterranean Problem," *Geog. Rev.* [476], 28
(1938), 83–101. [925]
Review of geog. literature, particularly with ref. to political geog.

ASIA (EXCLUDING U.S.S.R.)

(See also [88–91], guidebooks; [98], topographic mapping.)

Geomorphology: Spreitzer [214], 1938, pp. 243–247.—Soils: Giesecke
[230], 1939, pp. 250–266.—Economic geography: Lütgens [284],
1936, pp. 90–119.—Political geography: Vogel [318], 1934, pp. 269–
286. [926]

Mezhov", V. Ī. *Bibliografiia Aziī: Bibliographia Asiatica:* Series 2:
*Bibliographie des livres et articles des journaux russes concernant
l'Asie, la Sibérie exceptée.* St. Petersburg, 1891–1894. 3 vols.
666 pp. [927]
15,290 entries. In Vol. 3, Russian and foreign literature relating to the
peoples of Siberia is listed alphabetically, 882 entries.

Periodicals (See also [1174].)

The Journal of the Royal Asiatic Society
London. 1834–. Quarterly. [928]
Though largely devoted to hist. and philological subjects, contains also
much geog. material. Extensive notices of books and lists of accessions
to the Society's Library.
Branches of the Royal Asiatic Society have been established at Bom-
bay, at Colombo (Ceylon Branch), at Shanghai (North China Branch),
at Singapore (Malayan Branch), and at Seoul (Korea Branch). These
branches have published periodicals dealing with their respective regions.

Journal of the Royal Central Asian Society
London. 1914–. Quarterly. Prior to 1931 *Journal of the Central
Asian Society.* [929]
*Journal asiatique, ou recueil de mémoires, d'extraits, et de notices
relatifs à l'histoire, à la philosophie, aux langues, et à la littérature
des peuples orientaux*

Paris, Société Asiatique. 1821–[1938].   [930]
Contains material on the historical geog. of the Far East and of the
Islamic world.

Southwestern Asia

Weiss, Jakob. "Länder- und Völkerkunde des alten Orients: Vorderasien (ohne Kleinasien), 1911–34," *Geog. Jhrb.* [6], 50 (1935), 47–134.   [931]
By topics, with a section on regions. Primarily, though not exclusively, historical and archeological; 1,196 entries; ann.; auth. index.

Keller, Alexandre. "Bibliographie géologique et géographique de la Syrie, du Liban et des régions limitrophes." *Revue de géographie physique et de géologie dynamique* [218], 6 (1933), 453–512. [932]
By major regions, subcl. chron.; 829 entries.

Frey, Ulrich. "Vorderasien (1913–32)," *Geog. Jhrb.* [6], 47 (1932), 37–128.   [933]
Covers Asia west of India and south of the U.S.S.R. By regions, subcl. by topics, with a section on southwestern Asia as a whole; 1,582 entries; ann.; auth. index.

TURKEY

(See also [156], geographical studies.)

Mudge-Winchell [1]: bibliographies, d 67; commercial handbook (1926), a 150; statistical yearbook, a 129.   [934]

Fuller, Grace H. *Turkey: A Selected List of References.* (The Library of Congress, General Reference and Bibliography Division.) Washington. 1944. 114 mimeo. pp.   [935]
Emphasis on works published since 1923 in languages of western Europe and available in American libraries. By topics, incl. bibls., maps, geog., economic and social conditions; 916 entries; auth. index.

Migliorini, Elio. "La nuova Turchia: Viaggi e scritti recenti," *Bollettino della R. Società Geografica Italiana* [841], Series 7, 1 (1936), 589–621.   [936]
Review art. with bibl. refs. in footnotes; by topics, incl. bibls., travels, geology and physical geog., climate and vegetation, population, economics, political geog., regional studies; 204 entries; auth. index.

Periodical

Türk Coğrafya Dergisi [Turkish Geographical Review]
  Ankara, Türk Coğrafya Kurumu [Turkish Geographical Society]. 1943–. Quarterly. [937]

ARMENIA

Pratt, Ida A. "Armenia and the Armenians: A List of References in the New York Public Library," *Bull. of the New York Public Library,* 23 (1919), 123–143, 251–277, 303–335. Also published separately, with additions, New York, 1919. [938]
Compiled under the direction of Richard Gottheil. By topics, incl. bibls., periods., description and geog., geology and natural hist., hist., etc.; auth. index.

Periodical

Revue des études arméniennes
  Paris. 1920–. Irregularly. [939]
Though largely philological and historical, contains occasional material of geog. value, especially in the bibls.

IRAN

Wilson, Sir A. T. *A Bibliography of Persia.* Oxford, Clarendon Press, 1930. 263 pp. [940]
Alph. by auths.

Pratt, Ida A. "List of Works in the New York Public Library Relating to Persia," *Bull. of the New York Public Library,* 19 (1915), 9–126. [941]
By topics, incl. bibls., periods., geology and natural history, description and geog., hist., commerce, social life, science.

SYRIA

Notes by J. A. Tower on recent studies in Syrian geography, in *Geog. Rev.* [476], 27 (1937), 676–678; 29 (1939), 147–149. [942]

Masson, Paul. *Éléments d'une bibliographie française de la Syrie (géographie, ethnographie, histoire, archéologie, langues, littératures, religions).* (Chambre de Commerce de Marseille. Congrès

Français de la Syrie, Marseilles, les 3, 4, et 5 Janvier, 1919.) Marseille, 1919. 547 pp.   [943]
Deals exclusively with French works, listed chron.; 4,534 entries; auth. index; subj. index (system.).

PALESTINE

Mudge-Winchell [1]: statistical handbooks, a 127, b 22.   [944]

Thomsen, Peter. *Die Palästina-Literatur: Eine internationale Bibliographie in systematischer Ordnung, mit Autoren- und Sachregister.* Leipzig, Hinrichs. Vol. 1, for 1895–1904, published in 1911, 220 pp. (first issued in 1908 with title *Systematische Bibliographie der Palästina-Literatur*); Vol. 2, for 1905–1909, published in 1911, 336 pp.; Vol. 3, for 1910–1914, in 1916, 408 pp.; Vol. 4, for 1915–1924, in 1927, 774 pp.   [945]
By topics, incl. bibls., hist., archeology, historical geog., geog., modern Palestine, etc.; 8,235 entries (1927); genl. indexes.

Röhricht, Reinhold. *Bibliotheca geographica Palaestinae.* Berlin, 1890. 764 pp.   [946]
Covers period from A.D. 333 to 1878. Main bibl. in chron. order, with a section devoted to cartography; auth., place, and other indexes. For period 1878–1895 see bibls. accompanying each volume of *Zeitschrift des Deutschen Palästina-Vereins.*

Atlases

Smith, G. A. *Historical Atlas of the Holy Land.* 2d ed. London, Hodder and Stoughton, 1936. 60 pp. of maps; 21 pp. of text and index. 24 x 38 cm. Title of first edition (1915): *Atlas of the Historical Geography of the Holy Land.*   [947]
History (30 pp.); modern locational maps with hypsometric tints (22 pp.); also geology, vegetation, economic and political geog., missions, Jerusalem.

The Westminster Historical Atlas to the Bible. Ed. by G. E. Wright and F. V. Filson with an introductory article by W. F. Albright. Philadelphia, Westminster Press, 1945. 114 pp., 18 pls. of colored maps, 96 pp. of text, index, etc. 28 x 40 cm.   [948]

ARABIA

"List of Works in the New York Public Library Relating to Arabia and the Arabs, Arabic Philosophy, Science and Literature, *Bull. of*

the *New York Public Library,"* 15 (1911), 7-40, 163-198. Also published separately, New York, 1911. [949]
By topics, incl. bibls.; periods., hist., description and travel, science.

India and Burma

(See also [89], guidebooks; [269], medical geography.)

Mudge-Winchell [1]: geographical dictionary, a 325; government documents, a 373; historical reference books, a 357, b 51; statistical yearbooks, a 126.—Geographical work in India: notes by G. B. Cressey in *Geog. Rev.* [476], 34 (1945), 487-489, and by Shannon McCune in *Far Eastern Survey,* 15 (1946), 254-255. [950]

Cutts, E. H. "A Basic Bibliography for Indic Studies," *Amer. Council of Learned Societies Bull.,* No. 28 (May, 1939), pp. 445-505. [951]
By topics, incl. bibls., periods., hist., literature, philosophy, art, science, foreign contacts, etc.; 474 entries.—See also [153].

Cumming, Sir J. G. *Bibliography Relating to India (1900-1926).* (Royal Colonial Institute, London.) 1927. 16 pp. [952]
Short selective list of books in English. "Exclusively technical works and books having a purely local application have generally not been included." By topics, incl. geog., gazetteers, maps, politics, economics, hist., travel, etc.

Trinkler, Emil. "Vorderindien (1913-26)," *Geog. Jhrb.* [6], 42 (1927), 3-22. [953]
Covers India and Himalayas. Regional cl., subcl. by topics; 262 entries; ann.; auth. index.

Catalogue of the Library of the India Office. London, Vol. 1, 1888, 567 pp.; Vol. 1, Index, 1888, 207 pp.; Vol. 1, Supplement, 1895, 384 pp.; Vol. 1, Supplement 2, 1909, 619 pp. [954]
Continued from 1911 in a series of "Accessions" (see Besterman [3], Vol. 1, p. 491). By topics (incl. bibls., geog.) and places; genl. indexes.

———

Christian, J. L. *Modern Burma: A Survey of Political and Economic Development.* Los Angeles and Berkeley, Cal., Univ. of California Press, 1942. [955]
Bibl. on pp. 350-368.

Periodicals

The Indian Geographical Journal
    Madras, The Madras Geographical Association. 1926–. Quarterly.
    From 1926–1940 entitled *Journal of the Madras Geographical Association.*   [956]
Calcutta Geographical Review
    Calcutta Geographical Society. 1936–. Quarterly.   [957]
Bhugol: The Hindi Journal of Geography
    Allahabad. 1924–[1939]. Monthly.   [958]
    In Hindi.

Atlases

The Imperial Gazetteer of India. (Government of India.) Vol. 26.
Atlas. New (revised) ed. Oxford, 1931. 66 pl.; 48 pp. 14 x 22
cm.   [959]
Physical geog., population, languages, religions, agriculture, minerals,
transportation, hist., etc. (pl. 1–29); provincial maps (pl. 30–50; all but
2 on scale 1:4,000,000); town plans (pl. 51–60). Genl. index to provincial maps.

Crop Atlas of India. (Commercial Intelligence Department, India.)
Calcutta, Supt. Govt. Printing, India, 1923. 16 pl.; 2 pp. 21 x 34
cm.   [960]
Maps of India (160 miles to the inch) showing areas under cultivation
of 16 principal crops during the five years ending 1918–1919.

Eastern Asia: the Far East

(See also [269], medical geography.)

Mudge-Winchell [1]: commercial handbook (1932), a 150.   [961]

Pritchard, E. H., ed. *Far Eastern Bibliography.* 1936–.   [962]
The first 5 vols. covering 1936–1940 were published in mimeograph
form under the title *Bulletin of Far Eastern Bibliography* by the Committee on Far Eastern Studies of the American Council of Learned
Societies, Washington, D.C. The bibls. for 1941 and subsequent years
have appeared in the *Far Eastern Quarterly,* organ of the Far Eastern
Association. Topical and geog. cl.

Kerner, R. J. *Northeastern Asia: A Selected Bibliography: Contributions to the Bibliography of the Relations of China, Russia, and*

Japan, with Special Reference to Korea, Manchuria, Mongolia, and Eastern Siberia, in Oriental and European Languages. Berkeley, University of California Press, 1939. 2 vols. 714, 652 pp.   [963]
By countries, subcl. by topics (incl. bibls., periods., geog. and cartography, travels and explorations, hist., economics, international relations, etc.); 13,884 entries; subj. index (no auth. index).

Nunn, Janet H. *Bibliography on the Far East.* Washington, 1927. Mimeo. (U.S. Dept. of Commerce, Bureau of Foreign and Domestic Commerce, Division of Regional Information, *Special Circular,* No. 44 [–51].)   [964]
Lists of refs. to selected publs. in English (incl. govt. docs.) bearing on the Far East in general and particularly on trade and commerce. In 8 sections: 1, General bibliography on commerce and trade; 2. Far East; 3. Japan and Chosen; 4. China; 5. India, Burma and Ceylon; 6. Australia and New Zealand; 7. Philippine Islands, Hawaii and South Sea Islands; 8, Dutch East Indies, British Malaya, Siam and Indo-China. *Special Circular No. 213, 1930,* by the same compiler is entitled *A Selected Bibliography of the Far East.*

Bouterwek, Konrad. "Hinterindien und Indonesien, 1913–25," *Geog. Jhrb.* [6], 42 (1927), 22–86.   [965]
Covers Farther India, Malay Archipelago (except New Guinea), and the Philippine Islands. Place cl., subcl. by topics; 934 entries; ann.; auth. index. (See also [1165].)

Slade, W. A. "Bibliography of China, Japan, and the Philippine Islands," *Annals of the Amer. Acad. of Polit. and Social Science,* 122 (Nov., 1925), 214–246.   [966]
Lists books and periods. primarily in English; sections on (1) China, (2) Japan, (3) Philippine Islands, each subcl. by topics, incl. bibls., description and travel.

Periodicals

L'Asie française: Bulletin mensuel du Comité de l'Asie Française
Paris. 1901–[1940]. Monthly. Similar to *L'Afrique française* [1057].   [967]
A few reviews of books on Asiatic subjs. Mainly political in scope.

The Austral-Asiatic Bulletin
Melbourne, Australian Institute of International Affairs. 1937–[1942]. Bimonthly. [476]   [968]
See note in *Geog. Rev.* [476], 28 (1938), 336.

The Far Eastern Quarterly: Review of Eastern Asia and the Adjacent Pacific Islands
    Menasha, Wis., and New York, The Far Eastern Association. 1941–.
    Contains book reviews and bibls.   [969]

Bulletin de l'École Française d'Extrême-Orient
    Hanoi. 1901–[1937]. Index: 1901–20 in Vol. 21; 1921–30 in Vol. 32.   [970]

T'oung Pao, ou Archives concernant l'histoire, les langues, la géographie, et l'ethnographie de l'Asie orientale
    Leiden, E. J. Brill. 1890–[1940]. Normally 5 members a year.   [971]
    This period., edited since 1926 by the great French Orientalist Paul Pelliot, contains much bibl. material, incl. critical book reviews and notes.

Guidebooks

Guides Madrolle. Paris. Hachette.   [972]
    Vols. on North China (1911, 1913), South China, Java, Malaya, Siam, Indo-China, Philippines, Japan (1916), Indo China (1930).

Imperial Japanese Government Railways. *An Official Guide to Eastern Asia.* . . . 5 vols. Tokyo, 1913–1917.   [973]
    Vol. 1, Manchuria and Chosen; 2, Southwestern Japan; 3, Northeastern Japan; 4, China; 5, East Indies, incl. Philippine Islands, French Indo-China, Siam, Malaya, and Dutch East Indies.

CHINA

Mudge-Winchell [1]: bibliographies, a 394, c 80; commercial handbook (1926), a 150; encyclopedias, a 349–350; statistical yearbook, a 125, b 21, d 25.—Note by G. B. Cressey on some recent Chinese geographical studies, in *Geog. Rev.* [476], 26 (1936), 504–506. (See also *Geog. Rev.*, 27 (1937), 678–679; 35 (1945), 486–487.)   [974]

Quarterly Bulletin of Chinese Bibliography: English Edition. Published jointly by the Chinese National Committee on Intellectual Coöperation, Shanghai, and the National Library of Peiping. [1st ser.], 1934–1937, New series, 1940–.   [975]
    Current cl. lists of books in Chinese and in foreign languages, and of period. arts. Works on hist. and geog. are grouped together.

Conover, Helen F. *China: A Selected List of References on Contemporary Economic and Industrial Development with Special Emphasis on Post-War Reconstruction.* (The Library of Congress, General Reference and Bibliography Division.) Washington, 1945. 102 mimeo. pp.   [976]
By topics, incl. bibls., geog., agric., communications, mining, industry, commerce, social conditions; 924 entries; auth. and subj. indexes.

Chang, Chi-yun. "Geographic Research in China," *Annals of the Association of Amer. Geographers* [479], 34 (1944), 47–62.   [977]
Survey of recent progress in cartography, geophysics, geomorphology, climatology, soil geog., hydrography, oceanography, bio-, anthropo-, economic, political, historical, and regional geog., the hist. of geog., and geog. education, with bibl. refs. in the text and a short section on geog. societies.

National Library of Peiping. *Selected Chinese Books, 1933–1937.* Printed in China, 1939–1940. 214 pp.   [978]
Bibl. (cumulated from *Quarterly Bulletin of Chinese Bibliography* [975]) of books in Chinese on China. Authors' names and titles are given in Latin alphabet and in Chinese, with brief explanations of subj. matter in English. By topics, incl. bibls., philosophy, art, languages, geog. (pp. 107–126), hist., sociology, etc.; ann.; auth. index.

Hasenclever, Christa. "Bibliographie zur Industrialisierung Chinas," *Weltwirtschaftliches Archiv,* 45 (1937), 442–466.   [979]
By topics, incl. bibls., periods., genl. works, economics, individual industries; ann.

Robson, Harriet H. *Books for the Traveller or Sojourner in China.* New York, American Council, Inst. of Pacific Relations, 1937. 24 pp.   [980]
Covers "some of the best available books in English about various aspects of China." By topics, incl. maps, handbooks, geographies, culture, arts, language, literature; 138 entries; ann.; auth. index.

Gardner, C. S. *A Union List of Selected Western Books on China in American Libraries.* 2d ed. (American Council of Learned Societies.) Washington, D.C., 1938. 121 pp.   [981]
Aims "to include the greater number of those books which are now most valuable to serious students of China and things Chinese." Cites worth-while reviews and indicated holdings in American libraries. By topics (incl. bibls., periods., history, languages, etc.); 371 entries; auth. index.

Slade, W. A. "China: A Bibliographical List," *Annals of the Amer. Acad. of Polit. and Soc. Sci.,* 152 (1930), 378–398.   [982]
Supplements the China section of same author's *Bibliography of China, Japan, and the Philippine Islands* [966], 1925.

Köhler, Günther. "Ostasien, 1914–26: China nebst Mandschurei, Korea und Japan," *Geog. Jhrb.* [6], 42 (1927), 295–341.   [983]
Deals primarily with China, incl. Manchuria; short sections on East Asia as a whole and on Korea (none on Japan). Topical and place cl.; 760 entries; ann.; auth. index.

Cordier, Henri. *Bibliotheca Sinica: Dictionnaire bibliographique des ouvrages relatifs à l'Empire Chinois.* 2d ed. Paris, Guilmoto, 1904– 1908. 4 vols. xvi and 3,252 cols.   [984]
5 main parts: (1) China proper (incl. geog., ethnography and anthropology, climate and meteorology, natural hist., government, hist., religion); (2) foreigners in China; (3) relations between foreigners and Chinese; (4) the Chinese among foreign peoples; (5) countries tributary to China (i.e. Tartary, Tibet, Korea, Ryu Kyu Islands). The section devoted to geog. in Part 1 (248 cols.) is cl. by topics and provinces.

———

Knoepfmacher, Hugo. "Outer Mongolia: A Selection of References," *Bull. New York Public Library,* 48 (1944), 791–801.   [985]

Hu Huan-yong and Chen-Kang Tung. *Books and Articles on Sinkiang (in Western Languages).* (Institute for the Promotion of Dr. Sun Yat-san's [*sic*] Industrial Plan and Dept. of Geography, National Central University.) 1943. 47 pp. (Dept. of Geog., Inst. of Science, Graduate School, National Central Univ., *Bull. B.,* No. 2.)   [986]
Alph. by auths.

Periodicals

Ti-li Hsüeh-pao: Journal of the Geographical Society of China
Chungking, formerly Nanking. 1934–. Quarterly.   [987]
"The National Geographical Society of China, established in 1934, is a national organization of the geographers of China. Its list of members also includes many other distinguished scientists interested in geography" (Chang [977], 1944, p. 62).

Ti-hsüeh Tsa-chin: The Geographical Journal
   Peiping, Chinese Geographical Society. 1910–1937. Quarterly.
                                                                          [988]
Fang-chih: The Geographical Review
   Nanking, National Central University, Department of Geology
   and Geography. 1928–[1936?]. Quarterly. Chinese title
   varies.   [989]
Ti-li Hsüeh Chi-K'an: The Quarterly Journal of Geography
   Canton, Geographical Department of Sun Yat-sen University.
   1933–1934.   [990]
   The American Geographical Society also has *Bulletin of the Geographical Department,* No. 1, 1937. At least 7 numbers of this *Bulletin* were
   published. (See *Geog. Rev.* [476], 25 (1935), 499–500.)

Maps and Atlases

Chinese maps and atlases: note in *Geog. Rev.* [476], 24 (1934), 497–
   498.   [991]

Herrmann, Albert. *Historical and Commercial Atlas of China.*
   Cambridge, Mass., Harvard University Press, 1935. 84 pp. maps,
   32 pp. text (contents, bibl., index, etc.). 20 x 33 cm. (*Harvard-Yenching Inst. Monograph Ser.,* Vol. 1.)   [992]
   Hist. (pp. 5–63), modern naps, covering cities, political divisions, races,
   languages, economic geog., Chinese abroad, etc. (pp. 64–84). See review by C. W. Bishop, *Geog. Rev.* [476], 27 (1937), 515–517.

Ting, V. K., W. H. Wong, and S. Y. Tseng, eds. *New Atlas of the Chinese Republic.* Shanghai, Shen Pao Press. 1934. 53 pl. 22 pp.
   text, 180 pp. index. 24 x 34 cm. In Chinese.   [993]
   Primarily sectional locational maps; also city plans and genl. maps of
   all China showing communications, climate, population, languages,
   minerals, agriculture. See also *Gazetteer of Chinese Place Names Based on the Index to V. K. Ting Atlas,* Washington, D.C., Army Map Service,
   1944, 229 pp.

Dingle, E. J., ed. *The New Atlas and Commercial Gazetteer of China: A Work Devoted to Its Geography and Resources and Economic and Commercial Development.* (The Far-Eastern Geographical
   Establishment.) 2d ed. Shanghai, North China Daily News and
   Herald, [1918]. 325 pp. 22 pl. Also 6 colored maps and 18 diagrams
   in text. 39 x 54 cm.   [994]

Locational maps of the provinces and outer territories of China with names in Chinese and Roman characters; genl. maps of productions, railways, and forestry. The text provides a wealth of information, incl. statistical data, of commercial interest.

MANCHURIA

Mudge-Winchell [1]: historical and geographical dictionary, b 51; statistical yearbooks, a 127. [995]

Fochler-Hauke, Gustav. "Die Mandschurei (mit Dschehol) (1927–36)," *Geog. Jhrb.* [6], 53, Pt. 1 (1938), 275–326. [996] Topical cl.; 600 entries; ann.; auth. index.

TIBET

Manen, Johan van. "A Contribution to the Bibliography of Tibet," *Journal and Proceedings of the Asiatic Society of Bengal* (New Series), 18, (1922), 445–525. [997] Deals with bibl. aids to study of both Western and Tibetan works on Tibet.

KOREA

Mudge-Winchell [1]: bibliographies, a 357–358.—On recent books on Korea: note in *Geog. Rev.* [476], 36 (1946), 327–328. [998]

Lautensach, Hermann. "Korea (1926–36)," *Geog. Jhrb.* [6], 53, Pt. 1 (1938), 255–274. [999] By topics, with a section on regions; 247 entries; ann.; auth. index.

Underwood, H. H. "A Partial Bibliography of Occidental Literature on Korea from Early Times to 1930," *Trans. of the Korea Branch of the Royal Asiatic Society,* 20 (1931), 17–185, i–xvi. [1000] By topics, incl. bibls., periods., culture, science, hist., travel and description, etc., 2,882 entries; auth. index; selected list (pp. 184–185) of 50 titles. See also supplement by E. and G. Gompertz in *Trans. Korea Branch, R. Asiatic Soc.,* 24 (1935), 21–48 (369 entries).

JAPAN

(See also [91], guidebook.)

Mudge-Winchell [1]: encyclopedia, a 49; geographical names, d 61; statistical yearbooks, a 126, b 22. [1001]

Hellman, Florence S. *The Japanese Empire: Industries and Transportation; A Selected List of References.* (The Library of Congress, Division of Bibliography.) Washington, 1943. 56 mimeo. pp. [1002] 2 main sections: (1) books and pamphlets (subcl. under 2 hdgs: (a) directories, handbooks, and periods., (b) genl.); (2) arts. in periods., cl. by period. titles (alph.); 598 entries; subj. index.

Borton, Hugh, Serge Elisséeff, and E. O. Reischauer. *A Selected List of Books and Articles on Japan in English, French, and German.* (American Council of Learned Societies, Committee on Japanese Studies.) Washington, 1940. 151 pp. [1003] By topics, incl. bibls., ref. works, periods., geog., hist., economics, government and politics, sociology, religion, language, literature, art, etc.; 842 entries; auth. and title index.

Schwind, Martin. "Gross-Japan (1927–38)," *Geog. Jhrb.* [6], 54 (1939), Pt. 2 (1940), 373–449. [1004] Sections on the Japanese Empire as a whole (topical cl.) and on regions; 962 entries; ann.; auth. index.

Nachod, Oskar, and others. *Bibliographie von Japan* [*1906–1935*]. Leipzig, Hiersemann, 1928–1937. 5 vols. Essentially a continuation of von Wenckstern [1007], Vol. 2, 1907. Vols. 1 and 2 covering 1906–1926 were published in 1928 (also published with English title, *A Bibliography of the Japanese Empire,* London, Goldston, 1928); Vol. 3, 1927–1929, was published in 1931; Vol. 4, 1930–1932 (ed. and enlarged by Hans Praesent), in 1935; Vol. 5, 1933–1935 (compiled by Hans Praesent and Wolf Haenisch), in 1937. [1005] Covers works in western languages (incl. Russian). By topics, incl. bibls., periods., hist., travel, culture, natural sciences, etc.; 25,376 entries; auth. indexes.

Cordier, Henri. *Bibliotheca Japonica: Dictionnaire bibliographique des ouvrages relatifs à l'Empire Japonais rangés par ordre chronologique jusqu'à 1870, suivi d'un appendice renfermant la liste alphabétique des principaux ouvrages parus de 1870 à 1912.* Paris, Imprimerie Nationale: Leroux, 1912. xii pp. and 762 cols. [1006] Not on a scale comparable to Cordier's *Bibliotheca Sinica* [984]. "I have merely taken up in this volume the work of Pagès [see von Wenckstern [1007]] published in 1859, and I have carried it on, with corrections and additions, to the Revolution of 1868, or rather to 1870" (Preface, p. vi). The appendix is relatively brief (cols. 617–712). Refs.

are given to period. arts. and to reviews of many of the items. Auth. index. The chron. arrangement and lack of subj. index restrict the utility of this work for purposes of geog. research.

Wenckstern, F. von. *A Bibliography of the Japanese Empire, Being a Classified List of All Books, Essays and Maps in European Languages Relating to Dai Nihon (Great Japan) Published in Europe, America, and in the East.* Vol. 1, Leiden, Brill, 1895, 352 pp.; Vol. 2 (title slightly different), Tokyo, Osaka, and Kyoto, The Maruzen Kabushiki Kaisha, 1907, 502 pp.   [1007]
Vol. 1 covers the period 1859–1893; Vol. 2 the period 1894 to the middle of 1906. Appended to Vol. 1 is a facsimile reproduction of Léon Pagès, *Bibliographie japonaise, ou catalogue des ouvrages relatifs au Japon qui ont été publiés depuis le xvᵉ siècle jusqu'à nos jours,* Paris, 1859, 68 pp. Appended to Vol. 2 is a supplement to Pagès, *Bibliographie japonaise* (additions and corrections), 28 pp., and a *Systematic List of the Literature in Swedish Language on the Empire of Japan* compiled by Miss Valfrid Palmgren, 21 pp.

Wenckstern's work has been severely criticized (see Cordier, *Bibliotheca Japonica* [1006], p. v) and should be used with care. It does not cover Russian publs. By topics, incl. bibls., periods., travel, ethnography, natural hist., topography and hydrography, physiography, hist., art, literature, etc.; auth. indexes; index to periodicals.

Periodicals

Japanese Journal of Geology and Geography
Tokyo, National Research Council of Japan. 1922–[1941]. Quarterly.   [1008]
In English. Original monographs and abstracts of papers published in Japan on geology and physical geog.

Chigaku-Zasshi: Journal of Geography
Tokyo Geographical Society. 1889–[1941]. Monthly.   [1009]
In Japanese, with English tables of contents.

Chirigaku Hyoran: Geographical Review of Japan
Tokyo, Association of Japanese Geographers, Geographical Institute, Faculty of Science, Imperial University. 1925–[?] (at least 10 vols. were published). Monthly.   [1010]
Chiri-Ronso: Geographical Bulletin
Kyoto, Geographical Institute, Imperial University. 1932–1936. 1 to 3 vols. a year.   [1011]
In Japanese. See *Geog. Rev.* [476], 24 (1934), 496.

Atlas

Fujita, Motoharu, ed. *New Atlas of Japan* (*Shin Nippon Zucho*). (War Department, Army Map Service, Corps of Engineers, U.S. Army.) Washington, 1943. 34 pl. 241 pp. 19 x 26 cm.   [1012]
Reproduction (without color), with translations of the editor's introduction (dated 1934) and table of contents, of a Japanese atlas that is "particularly valuable in its listings of geographical names." Primarily locational maps.

Southeastern Asia and East Indies

The Southeast Asia Institute (formerly East Indies Institute), 15 West 77th Street, New York 24, N.Y.) began issuing in 1945 a series of *Selected Bibliographies* in mimeograph form. These list books and articles: No. 1 on the religions of Southeast Asia, No. 2 on Farther India, and No. 3, works in English on the government, economy, sociology, education and political movements of Southeast Asia (i.e. Burma, Siam, Malaya, Indo-China, East Indies, and Philippine Islands).—See also [1165].   [1013]

Kennedy, Raymond. *Bibliography of Indonesian Peoples and Cultures.* New Haven, Yale Univ. Press, 1945. 212 pp. (*Yale Anthropological Studies,* Vol. 4.)   [1014]
"Represents a close approximation to complete coverage of all extant books and periodical articles concerning the peoples and cultures of Indonesia. . . . The main focus is . . . on anthropology and sociology, including ethnography, archaeology, linguistics, and studies of acculturation." Many works on geog., colonial administration, education, economics, and hist. are also included, together with "standard references on geology, botany, zoology, and kindred subjects." Place cl. with a section on genl. works; no index.

INDO-CHINA

Martineau [331], 1932, pp. 345-364   [1015]

Boudet, Paul, and Rémy Bourgeois. *Bibliographie de l'Indochine française, 1913-1926.* Hanoi, 1929, 356 pp.   [1016]
Supplements Cordier, *Bibliotheca Indosinica* [1017]. Part 1, by topics (alph.), with a few anns.; Part 2, by auths.

Cordier, Henri. *Bibliotheca Indosinica: Dictionnaire bibliographique des ouvrages relatifs à la péninsule indochinoise.* Vols. 1-4, Paris,

Imprimerie Nationale, 1913–1915, 3,030 cols.; [Vol. 5], *Index,* by
M. A. Roland-Cabaton, Paris, Van Oest, Leroux, 1932. 309 pp.
(*Publications de l'École Française d'Extrême-Orient,* Vols. 15–18
and 18 bis.) [1017]
Vol. 1 is devoted to Burma, Assam, Siam, and the Laos country; Vol. 2
to the Malay Peninsula; Vols. 3 and 4 to French Indo-China and Cam-
bodia. Under each of these broad regional divisions the material is arranged topically much as in Cordier's *Bibliotheca Sinica* [984], with
extensive sections devoted to geog. Auth. index.

Periodicals

(See also [970].)

Bulletin Économique de l'Indochine
  Hanoi-Haiphong, Gouvernement Général de l'Indochine. 1898–
  [1933]. 6 nos. a year. [1018]
  Original material on geog., forests, ethnography, commerce, etc.

Cahiers de la Société de Géographie de Hanoi
  1922–1932. Irregularly. [1019]

Atlas

Atlas de l'Indochine. (Service Géographique de l'Indo Chine.)
  [1930?]. 51 pl. 29 x 41 cm. [1020]
  5 administrative maps, 18 sectional locational maps, 1:1,000,000; also
  mineral resources, meteorology, ethnography, hist., etc., and town plans.

MALAYA AND BRITISH NORTH BORNEO

Hellman, Florence S. *British Malaya and British North Borneo: A
Bibliographical List.* (The Library of Congress, Division of Bibliography.) Washington, 1943. 103 mimeo. pp. [1021]
By places and by topics, incl. bibls., genl., economics, natural hist.,
people, etc.; 980 entries; auth. and subj. indexes.

Robson, J. H. M. *A Bibliography of Malaya: Also a Short List of
Books Relating to North Borneo and Sarawak.* Kuala Lumpur,
F.M.S., 1939. 48 pp. [1022]
A list "of books in English relating to Malaya, together with a selection
of pamphlets, reports and papers on various subjects also relating to
Malayan affairs." By topics, incl. geology, travel, topography, hist.,
natural hist., etc.

Mudge-Winchell [1]: bibliographies, a 408; gazetteer, a 326. [1023]

Griffin, A. P. C. *A List of Books (with References to Periodicals) on the Philippine Islands in the Library of Congress. With Chronological List of Maps in the Library of Congress by P. Lee Phillips.* (The Library of Congress.) Washington, 1903. 412 pp. [1024]
(1) List of books: by topics (alph.), incl. bibls., description, discovery, hist., natural hist., govt. docs., etc.; (2) list of period. arts. (chron.); (3) list of maps, etc. (chron.); auth. and subj. indexes covering (1) and (2), place index for (3).

Pardo de Tavera, T. H. *Biblioteca Filipina, ó sea catálogo razonado de todos los impresos, tanto insulares como extranjeros, relativos á la historia, la etnografía, . . . la geografía, la legislación, etc., de las Islas Filipinas, de Joló y Marianas.* (The Library of Congress and the Bureau of Insular Affairs.) Washington, 1903. 439 pp. [1025]
Based mainly on one of the largest private collections of Philippine literature. By auths. (alph.); ann.

Atlas

Census Atlas of the Philippines, constituting *Census of the Philippines, 1939,* Vol. 5. Manila, Commonwealth of the Philippines, Commission of the Census, 1941. 44 pl.; 95 pp. 33 x 47 cm. [1026]
25 "geographic [i.e. locational] maps" of the several provinces; also population, climate, economy and resources, reproductions of early maps of the Philippines. 79 pp. of the text consist of alph. tables of municipalities, islands, and lakes, giving areas. See review by M. J. Proudfoot, *Geog. Rev.* [476], 32 (1942), 170–171.

NETHERLANDS INDIES AND NEW GUINEA

(See also [344, 1140a], bibliographies.)

Mudge-Winchell [1]: bibliographies, historical, a 358; commercial handbook (1923), a 150; encyclopedia, a 358; statistical yearbook, a 127. [1027]

Netherlands East Indies: A Bibliography of Books Published after 1930, and Periodical Articles after 1932, Available in U.S. Libraries. (The Library of Congress, Reference Department.) Washington, 1945. 208 pp. [1028]

A supplement covering books published before 1930 is in preparation. "A record is given of the location of the more unusual publications." System. by topics, incl. the natural scis., human geog., hist., economics, religion, language, govt., law, etc., with a section on individual islands (alph.); auth. index, indexes to anonymous works and selected titles.

Conover, Helen F. *The Netherlands East Indies: A Selected List of References.* (The Library of Congress, Division of Bibliography.) Washington, 1942. 46 mimeo. pp. [1029]
Lists modern books and period. arts. in English, French, and German, but not in Dutch. By topics, incl. bibls., periods., travel, cartography, geog., hist., economics, the arts, etc.; 446 entries.

Encyclopaedie van Nederlandsch Indie. 2d ed. The Hague, Nijhoff. 8 vols., 1917–1939; Vol. 9, incomplete (1939–1940). [1030]
This extensive regional gazetteer includes many bibl. data. Vols. 5–9, 1922–[1940], have been issued with continuous pagination in pamphlet form, as supplements to the original work in 4 vols., 1917–1921.

Literatuur-overzicht over het Jaar . . . van de taal-, land-, en volkenkunde en geschiedenis van Nederlandsch-Indië. K. Instituut voor de Taal-, Land-, en Volkenkunde van Nederlandsch-Indië te 's-Gravenhage. 1936–[1940]. Annual. [1031]
Title varies slightly. Published by Nijhoff, The Hague. First two numbers compiled by F. H. Van Naerssen, second two by H. Van Meurs. Topical and areal cl.; auth. indexes.

Catalogus der boekwerken betreffende Nederlandsch-Indië, aanwezig in de bibliotheek van het Central Kantoor voor de Statistiek (bijgewerkt tot 18 november 1938). Batavia, 1938. 309 pp. [1032]

Lundy, F. A. "The Dutch East Indies: A Bibliographical Essay," *Pacific Historical Review,* 2, No. 3 (1933), 305–320. [1033]
Running discussion followed by a list of 50 items, incl. Netherlands trade bibls. and bibl. aids and handbooks pertaining to the Netherlands Indies.

Wichmann, Arthur. *Entdeckungsgeschichte von Neu-Guinea,* forming Vols. 1 and 2, Parts 1, 2, and 3, of *Nova Guinea: Uitkomsten der Nederlandsche Nieuw-Guinea-Expeditie in 1903.* Leiden, Brill, 1909–1912. 2 vols. (in 4). [1034]
Essentially a bibl. of virtually all that had been published about New

Guinea, arranged in the order of the original explorers' expeditions. Genl. indexes in each vol.

Periodicals

Tijdschrift voor Indische Taal-, Land-, en Volkenkunde
Batavia, Bataviaasch Genootschap van Kunsten en Wetenschappen. 1855–[1942]. Recently quarterly.   [1035]
Bijdragen tot de Taal-, Land-, en Volkenkunde van Nederlandsch Indië
The Hague, Koninklijk Instituut voor de Taal-, Land-, en Volkenkunde van Nederlandsch Indië. 1853–[1941]. Normally monthly. Indexes (published by Martinus Nijhoff, The Hague): *Register*, 1853–1899; *Inhoudsopgave*, 1853–1933, 1934–1941. [1036]
Jaarverslag van den Topographischen Dienst in Nederlandsch-Indië
Batavia. 1905–[1936].   [1037]
Material on the topography and cartography of the Netherlands East Indies.

De Indische Gids

Amsterdam. 1879–[1932]. Monthly. Index, 1879–1903.   [1038]
Section devoted to book reviews and periodical literature.

AFRICA

(See also [98], topographic mapping; [181], historical geography.)

Mudge-Winchell [1]: commercial handbook, a 150; statistical yearbook, a 128.—Lewin [337], Vol. 1, 1930, pp. 1–37.—Geomorphology: Spreitzer [214], 1938, pp. 247–248.—Soils: Giesecke [230], 1939, pp. 266–272.—Economic geography: Lütgens [284], 1936, pp. 126–160.—Political geography: Vogel [318], 1934, pp. 288–293.   [1039]

Lewin, Evans. *Annotated Bibliography of Recent Publications on Africa, South of the Sahara, with Special Reference to Administrative, Political, Economic, and Sociological Problems.* London, 1943. 104 pp. (*Royal Empire Society Bibliography*, No. 9.)   [1040]
By topics, incl. bibls., administration, politics, economics, sociology; a few anns.; auth. index.

Worthington, E. B. *Science in Africa: A Review of Scientific Research*

Relating to Tropical and Southern Africa. London, New York, Oxford University Press, 1938.   [1041]
Bibl. on pp. 626–691: by topics, incl. surveys and maps, the various natural sciences, forestry, fisheries, agriculture, anthropology, health and medicine.

Weiss, Jakob. "Länder- und Völkerkunde des alten Orients: Nordostafrika (1911–36)," *Geog. Jhrb.* [6], 53, Pt. 1 (1938), 327–352.   [1042]
Primarily, though not exclusively, historical and archeological. Place cl., subcl. by topics; 312 entries; ann.; auth. index.

Kloss, H. "Schrifttum zur Kolonialfrage in Afrika," *Auslandkundliche Vorträge der Technischen Hochschule Stuttgart,* 8–9 (1934), 148–166.   [1043]
Covers bibls., periods., comprehensive works on geog., hist., etc. 3 main sections: (1) Africa as a whole, (2) former German colonies, (3) other colonies; (2) and (3) subcl. by colonies.

Work, M. N. *A Bibliography of the Negro in Africa and America.* New York, H. W. Wilson, 1928. 719 pp.   [1044]
Part I, "The Negro in Africa," pp. 1–247: by topics, incl. discovery and exploration, African civilizations, peoples, art, music, European governments and colonization, race problems, missions, etc.; ann.; auth. index. "A Bibliography of Bibliographies on Africa," pp. 242–247. See also [415].

Jaeger, Fritz. "Afrika, 1922–27," *Geog. Jhrb.* [6], 43 (1928), 3–41.   [1045]
By regions, subcl. by topics; 1,003 entries; ann.; auth. index.

Catalogue de la Bibliothèque [du Ministère des Colonies]. (Royaume de Belgique, Ministère des Colonies.) N.p., 1913. 498 pp.   [1046]
Covers Africa as a whole, but with especial emphasis on Belgian Congo. By topics, incl. bibls., hist., physical geog., economic geog., cartography, ethnography, missions, law, languages, medicine, etc.; 2,098 entries; auth. index.

Periodicals

African Affairs
London and New York, 1901–. Quarterly with special supplements issued irregularly. Prior to 1944: *Journal of the Royal African Society.*   [1047]
Includes much material of geog. value and critical reviews.

African World and Cape-Cairo Express
London, 1902–. Weekly.   [1048]
Contains many items of geog. interest.

Africa
London, International Institute of African Languages and Cultures. 1928–. Quarterly. Not published 1941–42.   [1049]
Contributions in English, French and German, of which many are of geog. value. Critical reviews and, in each issue, "bibliography of current literature dealing with African languages and cultures."

Journal de la Société des Africanistes
Paris, 1931–1938. Biennial.   [1050]
Contains arts. of geog. interest and an annual "Bibliographie africaniste" compiled by P. Lester.

Colonial Possessions of European Nations

Mudge-Winchell [1]: encyclopedia on former German colonies, a 348.—Lewin [337], Vol. 1, 1930, pp. 565–573 (Portuguese possessions); pp. 573–575 (Spanish possessions).   [1051]

BRITISH POSSESSIONS

Lewin [337], Vol. 1, 1930, pp. 76–454.   [1052]

Conover, Helen F. *The British Empire in Africa: A Selected List of References. I. General.* (The Library of Congress, Division of Bibliography.) Washington, 1942. 37 mimeo. pp.   [1053]
By topics, incl. bibls., periods., geog., hist., politics and government, economics, people; 358 entries; genl. index.

Conover, Helen F. *British East and Central Africa: A Selected List of References.* (The Library of Congress, Division of Bibliography.) Washington, 1942. 53 mimeo. pp.   [1054]
By topics (subcl. by places), incl. official publs., hist., description, travel, etc., physical aspects, politics and government, economics, people; 568 entries; genl. index.

FRENCH POSSESSIONS

Lewin [337], Vol. 1, 1930, pp. 475–527.   [1055]

Conover, Helen F. *French Colonies in Africa: A List of References.*

(The Library of Congress, Division of Bibliography.) Washington, 1942. 89 pp.   [1056]
By regions (with topical and place subcl.), incl. bibls., geog., hist., people, politics, economics, etc.; 1,265 entries.

Periodical

L'Afrique française: Bulletin mensuel du Comité de l'Afrique Française
     Paris. 1891–[1940].   [1057]
With its supplement, *Renseignements coloniaux,* this periodical, though devoted in the main to political and economic questions, contains materials of distinctly geog. importance in relation to the French possessions in Africa.

ITALIAN POSSESSIONS

Lewin [337], Vol. 1, 1930, pp. 554–564.—Martineau [331], 1932, pp. 558–613.   [1058]

Varley, D. H. *A Bibliography of Italian Colonisation in Africa with a Section on Abyssinia.* (Royal Empire Society and the Royal Institute of International Affairs.) London, 1936. 92 pp. (*Royal Empire Society Bibliography,* No. 7.)   [1059]
2 main parts: (1) Italian colonization in Africa, and (2) Abyssinia, each cl. by topics, incl. bibls., periods., maps, hist., administration, politics, agriculture, economics, etc.

Raccolta di pubblicazioni coloniali italiane. (Ministero degli Affari Esteri, Direzione Centrale degli Affari Coloniali.) Rome, 1911. 366 pp.   [1060]
3 main parts: (1) publs. on East Africa in genl. and on Ethiopia and Eritrea (by auths.); (2) publs. on Italian Somaliland and adjacent regions (by auths.); (3) bibl. cl. by topics, incl. bibls., periods., explorations, language, hist., natural hist., etc.; 1,378 entries in (1) and (2).

Periodicals

Africa italiana: Pubblicazione mensile dell'Istituto Fascista dell' Africa Italiana
     Rome, 1938–[1941]. Monthly. Succeeded *Revista coloniale: Organo dell'Istituto Coloniale Italiano,* 1906–1927; *L'oltremare,* 1927–1934; *Rivista delle colonie,* 1935–1938.   [1061]
Africa: Periodico mensile della Società Africana d'Italia

Naples. 1882–[1938]. Monthly. Title: *Africa, 1882–1885; Bolletino della Società Africana d'Italia,* 1886–1912; *L'Africa italiana: Bolletino della Società Africana d'Italia,* 1913–1937. [1062]
Contains book reviews and bibl. refs.

Atlas

Baratta, Mario, and Luigi Visintin. *Atlante delle colonie italiane, con notizie geografiche et economiche.* Novara, Instituto Geografico de Agostini, 1928. 36 colored pl.; 43 pp. text; 60 pp. photographs. 17 x 27 cm. [1063]
Predominantly locational maps with hypsometric tints. See review, *Geog. Rev.* [476], 19 (1929), 518.

Northwestern Africa

Martineau [331], 1932, pp. 267–310. [1064]

L'Afrique française du nord: Bibliographie militaire des ouvrages française ou traduits en français et des articles des principales revues françaises relatifs à l'Algérie, à la Tunisie et au Maroc [1830–1927]. (Ministère de la Guerre: État-Major de l'Armée: Service Historique.) Paris, Imprimerie nationale, 1930, 1935. 4 vols. [1065]
Vols. 1 and 2 deal with Morocco and Algeria, 3 and 4 with Tunisia; each cl. by topics, incl. bibls., hist., army, colonization, penetration of Sahara, etc.; 9,446 entries; ann.; no auth. index.

Periodical

. *Bulletin de la Société de Géographie d'Alger et de l'Afrique du Nord* Algiers. 1896–. Quarterly; annual, 1915–21. Indexes: 1896–1905, 1906–1922, 1923–1932 (*Bull.* No. 136, Trim. 4, 1933). [1066]
Contains book reviews and bibls. See especially Augustin Bernard, "Revue bibliographique des travaux sur la géographie de l'Afrique septentrionale," *Bulletin,* 1898–1903.

Atlas

Bernard, Augustin, and R. de Flotte de Roquevaire. *Atlas d'Algérie et de Tunisie.* (Gouvernement Général de l'Algérie: Direction de l'Agriculture, du Commerce et de la Colonisation: Service Cartographique.) Algiers (Carbonel) and Paris (Larose), 1923–. To be completed in 16 sections (pl. with text; many maps in text). 53 x 63 cm. [1067]

The principal maps are in color, scale 1:1,500,000. To date sections have appeared covering geology, hypsometry, climate, vegetation, demography, native population and rural habitations, official colonization, and agriculture. Sections pertaining to administrative divisions, soils, water power, rural land tenure, mineral resources, fisheries, commerce, etc., announced as in preparation.

MOROCCO

Martineau [331], 1932, pp. 294–310.   [1068]

Cenival, Pierre de, Christian Funck-Brentano, and Marcel Bousser. *Bibliographie marocaine 1923–1933.* Paris, Larose, 1937[?]. 606 pp.   [1069]
The annual bibls. published in *Hespéris* [1073] reissued in 1 vol.; subj. index. Continued 1934–1939 in *Hespéris,* 1939, pp. 321–389; 1943, pp. 3–122.

Lebel, Roland. *Les Voyageurs français du Maroc: L'Exotisme marocain dans la littérature de voyage.* Paris, 1936. 406 pp. (*Bibliothèque de culture et de vulgarisation nord-africaines.*)   [1070]
Covers per. 1556–1935; chron. list of works cited, pp. 395–403.

Bauer y Landauer, Ignacio. *Biblioteca hispano-marroqui: Apuntes para una bibliografia de Marruecos.* Madrid, Ed. Ibero-africano-americana, 1922. 1040 pp.   [1071]

Playfair, R. L., and Robert Brown. "A Bibliography of Morocco from the Earliest Times to the End of 1891," *Supplementary Papers, Royal Geographical Society,* Vol. 3, pp. 201–476. London, Murray, 1892.   [1072]
Chron. by titles; 2,243 entries; ann.; subj. and auth. indexes.

Periodicals

Hespéris: Archives berbères et Bulletin de l'Institut des Hautes-Études Marocaines
Paris. 1921–[1930]. Quarterly. Succeeds *Archives berbères,* 1915–1920; *Bulletin de l'Institut des Hautes-Études Marocaines,* 1920. Index to publications of the Institut des Hautes-Études Marocaines, 1915–1935, forms supplement to *Hespéris,* 1936.   [1073]
Revue de géographie marocaine
Casablanca, 1926–. Quarterly. Succeeds *Bulletin de la Société de*

Géographie du Maroc, 1916–1925.   [1074]
Bibl. notes.

Atlas

Lévi-Provençal, E., ed. *Maroc: Atlas historique, géographique et
économique.* Paris, Horizons de France, 1935. 95 pp., with 16 colored
maps in text. 27 x 32 cm.   [1075]
Mostly descriptive text with illustrations; the maps pertain to relief,
structure, hist., rainfall, languages, density of population, communications, agriculture, minerals, water power, tourism, etc.

ALGERIA

Martineau [331], 1932, pp. 274–284.   [1076]

Tailliart, Charles. *L'Algérie dans la littérature française: Essai de
bibliographie méthodique et raisonnée jusqu'à l'année 1924.* Paris,
E. Champion, 1925. 471 pp.   [1077]
By topics, incl. bibls., literature, geog., guidebooks, travels (subcl. regionally), hist., politics, etc.; 3,177 entries; detailed anns.; auth. and
subj. indexes.

Playfair, R. L. "A Bibliography of Algeria from the Expedition of
Charles V in 1541 to 1887," *Supplementary Papers, Royal Geographical Society,* Vol. 2, pp. 127–430. London, 1889. (Also reprint,
304 pp.)   [1078]
Chron. by titles; 4,745 entries; ann.; auth. and subj. indexes.

Playfair, R. L. *Supplement to the Bibliography of Algeria from the
Earliest Times to 1895.* London, Murray, 1898. 325 pp. (*Extra Publication, Royal Geographical Society.*)   [1079]

Periodical

*Bulletin trimestriel de la Société de Géographie et d'Archéologie
d'Oran*
    1878–. Quarterly. Indexes: 1878–1897; 1898–1907; 1908–
1927.   [1080]

TUNISIA

Martineau [331], 1932, pp. 285–294.   [1081]

Ashbee, H. S. *A Bibliography of Tunisia from the Earliest Times to*

the End of 1888. . . . London, Dulau, 1889. 144 pp.   [1082]
By auths.; genl. index.

Leconte, S., ed. *Tunisie: Atlas historique, géographique, économique,
et touristique.* Paris, Horizons de France, 1936. 107 pp., with 11
colored maps in text. 27 x 32 cm.   [1083]
Mostly descriptive text with illustrations; the maps illustrate natural
regions, vegetation, rainfall, Roman Tunisia, density of population,
minerals, agriculture, arts, and lines of communication.

WESTERN SAHARA

Monod, Th. "Notes bibliographiques sur le Sahara occidental," *Journal de la Société des Africanistes,* 3, 1933, 129–196. "Premier Supplément," *ibid.,* 3 (1933), 335–340.   [1084]
Les Territoires du sud de l'Algérie, Part 3, *Essai de bibliographie
(volumes, brochures, articles de revues, documents cartogra-
phiques).* (Gouvernement Général de l'Algérie, Commissariat
Général du Centenaire.) 2d ed. Algiers, Imprimerie Algérienne,
1930. 395 pp.   [1085]
By topics, incl. bibls., maps and surveys (93 pp.), geog., natural sciences,
hist., agriculture, public works, etc.; auth. index. The first edition
appeared in 1923.

Funck-Brentano, Christian. "Bibliographie du Sahara occidental,"
*Études, notes, et documents sur le Sahara occidental présentés au
VIIème Congrès de l'Institut des Hautes-Études Marocaines, Rabat,
30 Mai, 1930,* Rabat, 1930, pp. 203–296. Published also in *Hespéris*
[1073], 11 (1930), 203–296.   [1086]
By topics, incl. bibls., geog., cartography, natural hist., economics, ethnology, hist., politics, administration, etc.; 1,034 entries.

Periodical

Travaux de l'Institut de Recherches Sahariennes
Algiers, Université d'Alger. 1942–. Annual.   [1087]

Northeastern Africa

LIBYA

Mori, Attilio. *L'esplorazione geografica della Libia: Rassegna storica
e bibliografica.* Florence, 1927. 112 pp. (Governo della Cirenaica:

Ufficio Studi, *Rapporti e monografie coloniali,* Series 2ª, No. 5, December, 1926.) [1088]
Narrative and critical estimate of explorations in Libya with bibl. data in footnotes.

Ceccherini, Ugo. *Bibliografia della Libia* (*in continuazione alla "Bibliografia della Libia" di F. Minutilli*). (Ministero delle Colonie.) Rome, 1915. 213 pp. [1089]
By topics, incl. bibls., periods., maps, geog., hist., economics, etc.; 3,041 entries; auth. index.

Minutilli, Federico. *Bibliografia della Libia: Catalogo alfabetico e metodico di tutte le pubblicazioni . . . esistenti sino a tutto il 1902 sulla Tripolitania, la Cirenaica, il Fezzan e le confinanti regioni del deserto.* Turin, Bocca, 1903. 144 pp. [1090]
By topics, incl. bibls., geog., maps, hist., economics, politics, etc.; 1,269 entries; subj. and auth. indexes.

Playfair, R. L. "The Bibliography of the Barbary States, Part I: Tripoli and the Cyrenaica," *Supplementary Papers, Royal Geographical Society,* Vol. 2, pp. 557–614. London, 1889. [1090a]
Chron.; 579 entries; ann.; subj. and auth. indexes.

Periodicals

Bollettino geografico
Tripoli, Governo della Tripolitania. Ufficio Studi. 1931–1936. Biennially. [1091]
Archivio bibliografico coloniale (*Libia*)
Florence, Società Italiana per lo Studio della Libia e delle altre Colonie. 1915–July, 1921. Quarterly. In 1921 published by Istituto Agricola Coloniale Italiano: Sezione per lo Studio delle Colonie. [1092]
Detailed studies of the bibl. of the Italian colonies, more especially of Libya.

EGYPT AND ANGLO-EGYPTIAN SUDAN

Mudge-Winchell [1]: bibliography, a 350, d 64; geographical names, a 328. [1093]

Pratt, Ida A. *Modern Egypt: A List of References to Material in the New York Public Library.* New York, The New York Public Library, 1929. 320 pp. (Reprinted from *Bull. of the New York Public Library,* Vols. 32–33, 1928–1929.) [1094]

By many topics, incl. bibls., periods., climate, geog., hist., literature, economics, etc.

Lorin, Henri, ed. *Bibliographie géographique de l'Egypte.* (Société Royale de Géographie d'Egypte.) Cairo, 1928–1929. 2 vols. Vol. 1, *Géographie physique et géographie humaine,* by Henriette Agrel, Georges Hug, Jean Lozach, and René Morin, 488 pp. Vol. 2, *Géographie historique,* by Henri Munier, 271 pp.   [1095]
Vol. 1, topical cl., incl. bibls., travels, genl., math. geog. and cartography (subcl. by places), physical geog., anthropology and ethnology, human geog. (subcl. under hdgs. social geog., econ. geog.); 6,158 entries. Vol. 2, by major chron. pers.—(1) prehistoric, (2) Pharaonic, Greco-Roman, Coptic, and Byzantine, (3) Moslem to 1798—each subcl. by topics; 2,683 entries. Index to auths. and other proper names in each vol.

Pratt, Ida A. *Ancient Egypt: Sources of Information in the New York Public Library.* New York, The New York Public Library, 1925. 501 pp. (Reprinted from *Bull. of the New York Public Library,* Vols. 27–28, 1923–1924.) Continuation: I. A. Pratt: "Ancient Egypt, 1925–1941," *Bull. of the N.Y. Public Library,* Vols. 45–46 (1941–1942).   [1096]
By many topics, incl. bibls., periods., geog., hist., archeology, etc.

Gauthier, Henri. "Bibliographie des études de géographie historique egyptienne," *Bull. Soc. Sultanieh de Géographie* [1100], 9 (1919), 209–281.   [1097]
Review in chron. order of works on historical geog. of Egypt.

Sherborn, C. D. *Bibliography of Scientific and Technical Literature Relating to Egypt, 1800–1900.* (Ministry of Finance, Egypt: Survey Department.) Cairo, Govt. Press, 1915. 155 pp. (Reprint of Preliminary Edition, 1910.)   [1098]
Alph. by auths.; detailed subj. index.

———

Hill, R. L. *A Bibliography of the Anglo-Egyptian Sudan from the Earliest Times to 1937.* London, Oxford University Press, 1939. 224 pp.   [1099]
By topics, incl. anthropology, archaeology, arts, bibls., communications, forestry, geog. (subcl. by places), geology, hist., language, meteorology, zoology; auth. and topical indexes.

Periodical

Bulletin de la Société Royale de Géographie d'Égypte
  Cairo. 1875–. Irregularly. Until the First World War the society
was known as Société Khédiviale de Géographie; in 1915 it was
reorganized as the Société Sultanieh de Géographie, and in 1922
the designation was changed to the present one. Publication of a
new series of the *Bulletin* was begun in 1917. Index: 1875–
1927.   [1100]
Arts. in English and French. Bibls. and book reviews. In addition to the
Bulletin the Société Royal de Géographie d'Égypte has issued a number
of important geog. publs. These are published as *Mémoires* (16 issued
1919–1935), *Publications spéciales* (1924–), an unnumbered series in
which most of the publs. are historical studies, and a number of independent publs. not in a series.

Atlas

*Atlas of Egypt: A Series of Maps and Diagrams with a Descriptive
  Text Illustrating the Orography, Geology, Meteorology, and Economic Conditions.* Presented to the International Geographical
Congress at Cambridge, 1928, by Command of His Majesty King
Fouad I. (Egyptian Government.) Giza, Survey of Egypt, 1928.
31 pl. 81 x 70 cm.   [1101]
Locational maps with hypsometric tints; also climate, population, agriculture, etc. See review by Eugene Van Cleef, *Geog. Rev.* [476], 19
(1929), 342–343.

Western Africa

BRITISH WEST AFRICA

Conover, Helen F. *British West Africa: A Selected List of References.*
(The Library of Congress, Division of Bibliography.) Washington, 1942. 32 mimeo. pp.   [1102]
By topics (subcl. areally), incl. official publs., hist., description, geog.,
physical aspects, politics and government, economics, people; 333 entries; genl. index.

Cardinall, A. W. *A Bibliography of the Gold Coast.* Issued as a Companion Volume to the Census Report of 1931. Accra, Govt. Printer.
403 pp.   [1103]

By chron. pers. and by topics, incl. bibls., maps, climate, resources, etc.; 5,168 entries; auth. index. For many items fails to indicate author's initials, place of publication, etc.

Luke, H. C. *A Bibliography of Sierra Leone, Preceded by an Essay on the Origin, Character and Peoples of the Colony and Protectorate*. London, Oxford University Press, 1925. 230 pp.   [1104]
Chron. under hdgs. for genl. literature, native languages, colonial laws and ordinances, arts. in journals of societies and in periods., state and parliamentary papers, maps; 1,103 entries; topical and auth. indexes.

FRENCH WEST AFRICA

Martineau [331], 1932, pp. 333-344.   [1105]

Joucla, Edmond. *Bibliographie de l'Afrique Occidentale Française*. Avec la collaboration des services du Gouvernement Général de l'Afrique Occidentale Française et pour le Dahomey. Paris, Société d'éditions Géographiques, Maritimes et Coloniales, 1937. 704 pp. *(Bibliographie générale des colonies françaises,* par G. Grandidier [et] E. Joucla.)   [1106]
Alph. by auths., with hdgs. for place and tribal names (giving cross-refs. to auth. entries) and also 1,672 entries under hdg. "Cartes et croquis"; 9,543 entries; no anns. except under entries for maps; index to auths. and geog. names; ingenious subj. index consisting essentially of a series of separate genl. indexes for a wide variety of topics (incl. bibls., maps, geog., geology, hist., etc.).

Tuaillon, J. L. *Bibliographie critique de l'Afrique Occidentale Française*. Paris, Lavauzelle, 1936. 52 pp.   [1107]
Covers French publs. only: (1) selection of books and (2) periods. (topical cl., incl. hist., geog.; critical anns.); (3) supplementary lists of books, periods., and period. arts.

Urvoy, Y. "Essai de bibliographie des populations du Soudan central (Niger français, nord de la Nigeria anglaise)," *Bulletin du Comité d'Études Historiques et Scientifiques de l'Afrique Occidentale Française,* 19 (1936), 243-333.   [1108]
By chron. pers. and places, with a section on linguistics; 732 entries; a few anns.; auth. index.

Lebel, A. R. *L'Afrique occidentale dans la littérature française (depuis 1870)*. Paris, Larose, 1925. 290 pp.   [1109]
Not a bibl. but a broad discussion of the subj. with bibl. footnotes. 3

main parts: (1) literature of voyages and conquest, (2) technical literature, (3) imaginative literature. List (chron.) of French works on French West Africa; auth. index.

Periodical

Bulletin du Comité d'Études Historiques et Scientifiques de l'Afrique Occidentale Française
  Paris. 1918–[1938]. Quarterly. Succeeds *Annuaire et mémoires* of same Comité, published at Gorée, 1916–1917. [1110]
  Contains many papers of geog. interest.

Maps and Atlases

Martonne, Ed. de. *Inventaire méthodique des cartes et croquis, imprimés et manuscrits, relatifs à l'Afrique occidentale existant au gouvernement général de l'A.O.F. à Dakar.* (Gouvernement Général de l'Afrique Occidentale Française: Service Géographique.) Laval, Goupil, 1926. 154 pp. [1111]
  Place cl. with section on genl. maps, hydrographic maps, and maps of colonial boundaries; index to auths. and geog. names.

———

Atlas des cercles de l'Afrique Occidentale Française. Dressé et dessiné au Service Géographique de l'A.O.F., à Dakar, sous la direction du Commandant Ed. de Martonne. . . . (Gouvernement Générale de l'Afrique Occidentale Française.) Paris, Forest, 1924–1926. 8 vols. (fascicules). 115 pl. 34 x 48 cm. [1112]
  A map for each *cercle* (administrative division) showing boundaries, drainage, towns and villages, and routes, but not relief.

LIBERIA

Lewin [337], Vol. 1, 1930, pp. 575–577. [1113]

ANGOLA

Borchardt, Paul. *Bibliographie d'Angola (Bibliotheca Angolensis), 1500–1910.* (Instituts Solvay: Institut de Sociologie.) Brussels, etc., 1912. 65 pp. (*Monographies bibliographiques, publiée par l'Intermédiaire Sociologique,* No. 11.) [1114]
  By topics, incl. historical, physical, biological and economic geog., and cartography.

Central Africa

FRENCH EQUATORIAL AFRICA

Martineau [331], 1932, pp. 311–332. [1115]

Bruel, Georges. *Bibliographie de l'Afrique Équatoriale Française.* (Gouvernement Général de l'Afrique Équatoriale Française.) Paris, E. Larose, 1914. 330 pp. [1116]
2 parts: (1) books and period. arts. (alph. by auths.), (2) list of anonymous arts. in periods. (alph. by period. titles); 7,029 entries; subj. index (of questionable utility).

BELGIAN CONGO

Martineau [331], 1932, pp. 9–85.—Lewin [337], Vol. 1, 1930, pp. 536–553. [1117]

Simar, T. "Bibliographie congolaise de 1895 à 1910," *La Revue congolaise,* 2 (1912), 252–283, 354–381. [1118]
Continuation of Wauters [1119], 1895. Not a bibl., strictly speaking, but a "systematic guide for those who wish to undertake studies of the Congo. A publication useful in its modesty" (*Ann. de Géog., Bibl.* [7], 22 (1912), 252). Topical cl.

Wauters, A.-J. *Bibliographie du Congo, 1880–1895: Catalogue méthodique de 3,800 ouvrages, brochures, notices et cartes relatifs à l'histoire, à la géographie et à la colonisation du Congo.* Brussels, Administration du Mouvement Géographique, 1895. 405 pp. [1119]
By topics, incl. periods., maps, hist., exploration, physical geog., flora, fauna, anthropology, languages, missions, economic conditions, etc.; index to auths. and travelers' names. Introductory comment precedes each section, indicating the outstanding works.

Periodicals

Congo: Revue générale de la colonie belge
Brussels. 1920–[1940]. 10 numbers a year. Succeeds to *La Revue congolaise* (1910–1914). Index: 1920–25. [1120]
Arts., book reviews, and refs. to period. arts. on Congo.

Bulletin de la Société Belge d'Études Coloniales
Brussels. 1894–1925. Monthly. Merged with *Congo* [1120] in 1926. [1121]
Contains book reviews.

Atlas

Rouck, René de. *Atlas géographique et historique du Congo belge et des territoires sous mandat du Ruanda-Urundi.* Brussels, Rouck, 1938. 12 pl. 13 pp. Index. 26 x 35 cm. [1122]
Locational maps with hypsometric tints of Belgian Congo, 1:3,000,000; also climate, geology, hist., boundaries, etc.

Eastern Africa

Migliorini, Elio. "Elenco di scritti d'interesse geografico relativi all' Africa Orientale Italiana pubblicati negli anni 1935 e 1936 [and 1937–1938]," *Bollettino della R. Società Geografica Italiana* [841] Ser. 7, Vol. 2 (1937), 661–697; Vol. 4 (1939), 883–911. [1123]
By topics, incl. bibls., maps, natural hist., hist., population, exploration, colonization, etc.

ETHIOPIA AND ERITREA

Lewin [337], Vol. 1, 1930, pp. 577–582. [1124]

Thomas, T. H. "Modern Abyssinia: A Selected Geographical Bibliography," *Geog. Rev.* [476], 27 (1937), 120–128. [1125]
Deals with 40 recent geog. publs.

Papieri, Mario. *Contributo alla bibliografia e cartografia della Somalia Italiana.* Rome, Istituto Coloniale Fascista, 1932. 90 pp. [1126]

Fumagalli, Giuseppe. *Bibliografia etiopica: Catalogo descrittivo e ragionato degli scritti pubblicati dalla invenzione della stampa fino a tutto il 1891 intorno alla Etiopia e regioni limitrofe.* Milan, Ulrico Hoepli, 1893. 299 pp. [1127]
By topics, incl. bibls., maps, language, literature, physical geog., hist., natural hist., etc.; 2,758 entries; a few anns.; auth. index. See also Silvio Zanutto: *Bibliografia etiopica in continuazione alla "Bibliografia etiopica" di G. Fiumagalli: Primo contributo, bibliografia,* Rome, Sindicato Italiano Arti Grafiche, 1929. 36 pp. (2d ed., 1936?, 56 pp.).

Dainelli, Giotto, Olinto Marinelli, and Attilio Mori. "Bibliografia geografica della colonia Eritrea, I: anni 1891–1906," *Revista Geografica italiana* [843], 14 (1907). 72 pp. (separately paged at end of vol.). [1128]
Running account of progress of geog. investigation of Eritrea with bibl. data in the footnotes.

Madagascar and Other Islands of the Indian Ocean

Martineau [331], 1932, pp. 365–374 (Madagascar); pp. 406–408 (La Réunion, etc.).—Lewin [337], Vol. 1, 1930, pp. 523–526.   [1129]

Platt, Elizabeth T. "Madagascar: Great Isle, Red Isle: A Bibliographical Survey," *Geog. Rev.* [476], 27 (1937), 301–308.   [1130]
Brief survey of recently published material.

Grandidier, Guillaume. *Bibliographie de Madagascar*. 3 vols. with continuous pagination: (1) Part I, Paris, Comité de Madagascar, 1905, 441 pp.; (2) Part II, 1906, 479 pp.; (3) *Bibliographie de Madagascar, 1904–1933,* Paris, Société d'Éditions Géographiques, Maritimes et Coloniales, 1935, 591 pp. (*Bibliographie générale des colonies françaises.*)   [1131]
Extremely comprehensive. (1) and (3) contain lists of titles (alph. by auths.); (2) and (3) contain lists of administrative and political documents, manuscripts, periods., etc.; 14,978 entries; subj. index (incl. hdgs. for various branches of geog.) to whole work in (3).

Periodical

La Revue de Madagascar
Tananarive, Gouvernement Général. 1933–1939. Quarterly. Succeeds the section "Partie Documentation" of Colonie de Madagascar et Dependences: *Bulletin economique,* 1901–[1930]. [1132]
Arts. of geog. interest and a bibl. section.

South Africa and the Rhodesias

Mudge-Winchell [1]: atlas, historical, a 337; bibliographies, a 360, d 35; geographic names, a 331; statistical yearbooks, a 128. [1133]

Mendelssohn, Sidney. *South African Bibliography. . . .* London, Kegan Paul, 1910. 2 vols: Vol. 1, 1,080 pp.; Vol. 2, 1,139 pp. *Additions, 1910–1914,* 1914.   [1134]
Covers South Africa from the earliest times to the date of publ. An extensive and elaborate work. In the auth. catalogue (alph.) (Vol. 1, pp. 1–1008; Vol. 2, pp. 1–652) comments (some of them lengthy) are given upon the various authors and works listed. At end of Vol. 2: lists of (1) South African Imperial Blue-Books (by colonies, subcl.

chron.), (2) period. arts., (a) in South African periods. (by auths. under titles of periods.), (b) in other periods. (by auths. under place hdgs.). Detailed chron. and topographical index (topical and place cl.); index to maps.

Conover, Helen F. *British Empire in Africa, IV. The Union of South Africa: A Selected List of References.* (The Library of Congress, Division of Bibliography.) Washington, 1943. 77 mimeo. pp. [1135]
By topics, incl. bibls., hist., geog., politics and government, economics, people, with a section on Rhodesia; 760 entries; genl. index.

Schapera, Isaac, ed. *Select Bibliography of South African Native Life and Problems.* Compiled for the Inter-University Committee for African Studies. London, Oxford University Press, 1941. 261 pp.   [1136]
By topics under main hdgs. physical anthropology, archeology, ethnography (subcl. by ethnic groups), modern status and conditions, linguistics; ann.; auth. index.

"Works Relating to South Africa in the New York Public Library," *Bull. of the New York Public Library,* 3 (1899), 429–461. See also supplementary list on pp. 502–505.   [1137]
Place cl., with section on South Africa in genl. (incl. maps, hist., etc.).

Periodicals

The South African Journal of Science, Being the Report of the South African Association for the Advancement of Science
Johannesburg. 1903–. Annual.   [1138]
Contains substantial articles of geog. interest.

The South African Geographical Journal
Johannesburg, South African Geographical Society. 1917–. Annual.   [1139]
Arts. and reviews.

Atlas

Walker, E. A. *Historical Atlas of South Africa.* Oxford University Press, 1922. 26 pl.; 26 pp. 26 x 38 cm.   [1140]
In addition to maps illustrating the internal history of South Africa south of the Zambesi, includes maps showing "the connection between Africa and the East Indies from the fifteenth to the early nineteenth centuries," "the bearing on South African history of the European partition of Africa," and economic maps explaining that partition.

AUSTRALIA AND NEW ZEALAND

Lewin, Evans. *Best Books on Australia and New Zealand: An Annotated Bibliography.* London, 1946. 63 pp. (*Royal Empire Soc. Bibls.* [330], No. 13.)   [1140a]
Lists books published for the most part since 1900. Sections on (1) Australia, (2) New Zealand, and (3) New Guinea, Samoa, and other Island Dependencies; (1) and (2) subcl. by topics, incl. bibls., discovery and exploration, hist., economics, etc.

Australia

Mudge-Winchell [1]: bibliographies, a 384, c 76, d 72–73; commercial handbook (1922), a 150; encyclopedia, a 348; geographic names, a 327; government documents, a 372; statistical handbooks and yearbooks, a 124, c 26 (Queensland), d 24.—Lewin [337], Vol. 2, 1931, pp. 1–383.—Soils: Giesecke [230], 1939, pp. 272–274.—Economic geography: Lütgens [284], 1936, pp. 119–125.   [1141]
A bibliography of Australian geographical literature from 1926 has been published from time to time in the *Australian Geographer* [1147], beginning in 2 (1933), 22–28.   [1142]

Fuller, Grace H. *Australia: A Selected List of References* (The Library of Congress, Division of Bibliography.) Washington, 1942. 101 mimeo. pp.   [1143]
By topics, incl. bibls., description and travel, hist., economics, social and cultural life, natural hist., etc.; 793 entries; auth. index.

Geisler, Walter. "Australien und Ozeanien (1928–37)," *Geog. Jhrb.* [6], 53, Pt. 2 (1938), 566–630.   [1144]
By regions (incl. New Guinea), subcl. by topics; 1,049 entries; ann.; auth. index.

Australasian Bibliography: Catalogue of Books in the Free Public Library, Sydney, Relating to, or Published in, Australia. Sydney, 1893.   [1145]
In 3 parts: Part 1, 444 pp., by auths.; Part 2, 589 pp., by colonies, subcl. by auths.; Part 3, 229 pp., a series of classified subj. and title catalogues, incl. a section devoted to geog., topography, voyages, travels (place cl.).

Though now old and containing much material not geographical, this is a work of importance to the student of the hist. of the exploration of Australia.

Steere, F. G. *Bibliography of Books, Articles, and Pamphlets Dealing with Western Australia, Issued Since Its Discovery in 1616.* Perth, Simpson, 1923. 183 pp.   [1146]
By topics, incl. periods., maps, discovery and early voyages, travel and exploration, aboriginal inhabitants, natural hist., hist., government, etc.; auth. index.

Periodicals

Proceedings of the Royal Geographical Society of Australasia: South Australian Branch
Adelaide. 1885–. Annual.   [1147]

The Australian Geographer
Sydney, The Geographical Society of New South Wales. 1928–. Irregularly. Subtitle until 1933 *Journal of the Geographical Society of New South Wales.*   [1148]
Contains arts. written in the spirit of modern geog. Its initiation was due to the influence of Professor Griffith Taylor.

Victorian Geographical Journal: Including the Proceedings of the Royal Geographical Society of Australasia (Victoria)
Melbourne. 1912–1920. Irregularly. Succeeds *Transactions and Proceedings,* 1883–1911. Incorporated with *Victorian Historical Magazine* in 1921, Historical Society of Victoria, Melbourne. Index: 1883–1918 in Vol. 34.   [1149]

Queensland Geographical Journal (New Series), Including the Proceedings of the Royal Geographical Society of Australasia, Queensland
Brisbane. 1899–. Annual. Succeeds *Proceedings and Transactions of the Queensland Branch of the Royal Geographical Society of Australasia,* 1886–1898.   [1150]

Proceedings of the Royal Geographical Society of Australasia: New South Wales Branch
Sydney. 1885–1898. Irregularly.   [1151]

New Zealand

Mudge-Winchell [1]: gazetteer, a 326; geographic names, b 47; statistical works, a 127, c 26.—Lewin [337], Vol. 2, pp. 384–482.   [1152]

"Bibliography of Post-War Publications of Economic Interest on New Zealand," *Economic Record: Journ. of the Economic Society of Australia and New Zealand,* 15 (1939), Supplement, 158–170.   [1153]

Hocken, T. M. *A Bibliography of the Literature Relating to New Zealand.* Wellington, N.Z., 1909. 631 pp. *Supplement* . . . by A. H. Johnstone, Auckland, 1927, 75 pp.   [1154]
Chron.; ann.; detailed genl. index.

Periodical

The New Zealand Geographer
   Christchurch, New Zealand Geographical Society. 1945–. [1155]
   See note in *Geog. Rev.* [476], 35 (1945), 496.

ADDENDA

The following references were obtained too late for inclusion in the Bibliography, above. They are, however, numbered consecutively with the references in the Bibliography and indexed accordingly.

A Selected Bibliography of Geography and Allied Subjects Published in the United States between December 31, 1938, and July 1, 1946. College Park, Maryland, 1946. 19 mimeo. pp.   [1156]

Distributed by the American Society for Professional Geographers. Secretary: E. Willard Miller, Department of Geography, Pennsylvania State College, State College, Pennsylvania. Lists books alph. by auths.; brief anns.

Geographisches Jahrbuch [6]. The following bibliographical reviews appeared in Vols. 55 (1940), 56 (1941), 57 (1942), and 58 (1943), Pt. 1. The references to those in Vols. 55 and 57, Pt. 2, were furnished by courtesy of Dr. B. W. Adkinson, Acting Chief, Division of Maps, Library of Congress; the others were derived from copies received at the American Geographical Society in December, 1946.

Blüthgen, Joachim. "Dänemark (1930–40)." 56 (1941), Pt. 2, 267–307. 429 entries.   [1157]

Idem. "Finnland (1929–40)." 56 (1941), Pt. 2, 308–356. 487 entries.   [1158]

Idem. "Schweden (1929–40)." 56 (1941), Pt. 1, 27–114. 1000 entries.   [1159]

Dietrich, Bruno. "Nordamerika (1931–1942): Der Erdtail (Vereinigte Staaten von Amerika)." 58 (1943), Pt. 1, 87–232.   [1160]

In two sections, each cl. by topics: A) General North America; B) The United States. 2353 entries. (The concluding part (presumably on the United States by regions) was scheduled for 58 (1943), Pt. 2).

Dörries, H. "Siedlungs- und Bevölkerungsgeographie (1908–38)." 55 (1940), Pt. 1, 3–380. 4995 entries.   [1161]

Evers, W. "Norwegien (1929–40)." 56 (1941), Pt. 1, 4–26. 212 entries.   [1162]

George, Pierre. "Frankreich (1928–42)." 57 (1942), Pt. 2, 457–546. 1152 entries. [1163]
Haack, H. "Die Fortschritte der Kartographie (1936–1942)." 57 (1942), Pt. 1, 4–137, Pt. 2, 365–456. 2886 entries. [1164]
Helbig, Karl. "Hinter- und Insel-Indien (1926–1939/40)." 57 (1942), Pt. 1, 138–343; Pt. 2, 547–769. 3576 entries. [1165]
Herrmann, A. "Geschichte der Geographie (1926–39)." I) "Bis zum Ausgange des Mittelalters," 55 (1940), Pt. 1, 381–434; II) "Vom Zeitalter der Entdeckungen bis zum Ende des 19 Jahrhunderts," 55 (1940), Pt. 2, 433–543. 2733 entries. [1166]
Migliorini, Elio. "Italien (1931–39)." 56 (1941), Pt. 1, 115–194. 1172 entries. [1167]
Mikula, H. "Protektorat Böhmen und Mähren, Sudetenland, Slowakei, die Waldkarpathen und deren südliches Vorland (1928–38/39)." 55 (1940), Pt. 2, 463–534. 846 entries. [1168]
Morawetz, Sieghard. "Donau und Alpengaue (1929 bis Juli 1942)." 58 (1943), Pt. 1, 4–86. 1024 entries. [1169]
Tuckermann, Walther. "Kanada (und Neufundland) 1931–39." 56 (1941) Pt. 2, 357–432. 900 entries. [1170]
Vosseler, P. "Schweiz (1929–39)." 55 (1940), Pt. 2, 534–632. 989 entries. [1171]

The Journal of Glaciology

London, The British Glaciological Society, 1947–. [1172]

Pacific Science: A Quarterly Devoted to the Biological and Physical Sciences of the Pacific Region
Honolulu, The University of Hawaii, 1947–. [1173]

The Middle East Journal
Washington, The Middle East Institute, 1947–. Quarterly. [1174]

APPENDIX

CLASSIFIED INDEX OF AMERICAN PROFESSIONAL GEOGRAPHERS, LIBRARIES OF GEOGRAPHICAL UTILITY, AND INSTITUTIONS ENGAGED IN GEOGRAPHICAL RESEARCH

EXPLANATORY NOTE

DURING the spring and summer of 1945 a questionnaire was sent to the 425 members of the two nation-wide organizations of professional geographers, the Association of American Geographers and the American Society for Professional Geographers. To this questionnaire, 210 replies have been received. Each recipient was asked (1) to indicate the main branches of geography, or the areas, or both, with which his research work has dealt, and (2), for each subject so indicated, also to give the names of (a) libraries deemed to be particularly well equipped for purposes of geographical research, (b) institutions known to be actively engaged in such research, and (c) living persons qualified to give helpful counsel to others who might wish to conduct such research.

Each library and each institution mentioned by five or more respondents to the questionnaires in answer to questions (2a) and (2b) is listed below. The figures in parentheses show the number of respondents who mentioned each. While one may well hesitate to draw general conclusions from statistics of this kind, the first list would seem to give a clue to the relative amount of use made of the different libraries by a representative group of American geographers (compare the list on p. 65 above) and the second at least to suggest the relative degree of productivity in *geographical* research that the same group would ascribe to the different institutions as of 1945.

Libraries.—Library of Congress (85), American Geographical Society (53), Harvard Univ. (34), U.S. Dept. of Agriculture (agricultural geog., soils, climatology, land utilization, U.S.A.) (31), Univ. of Chicago (31), Univ. of California (26), Clark Univ. (23), Univ. of Michigan (23), New York Public Library (19), U.S. Geological Survey (geomorphology, U.S.A.) (17), Columbia Univ. (15), U.S. Weather Bureau (climatology) (15), Yale Univ. (14), Univ. of Wisconsin (13), Blue Hill Observatory (climatology) (10), John Crerar Library (9), U.S. Dept. of Commerce (economic geog.) (8), U.S. Dept. of State (8), Pan American Union (Latin America) (6), Univ. of Texas (6), Ohio State Univ. (6), Univ. of Illinois (5), Univ. of North Carolina (U.S.A.—the South) (5), Huntington Library (5).

Research Institutions.—U.S. Dept. of Agriculture (agricultural geog., soils, climatology, land utilization and planning, U.S.A.) (31), Univ. of California (24), Harvard Univ. (24), American Geographical Society (21), Univ. of Chicago (20), Clark Univ. (10), Univ. of Michigan (10), U.S. Geological Survey (geomorphology) (10), Univ. of Wisconsin (9), U.S. Weather Bureau (climatology) (8), Inst. of Pacific Relations (Pacific, Far East) (7), Smithsonian Instn. (6), Columbia Univ. (6), U.S. Office of Strategic Services * (cartography) (6), Blue Hill Observatory (climatology) (5).

The following index presents in detail a part of the information gathered through the questionnaire. Inquiries of this kind always yield data of uneven value. In this case, the information concerning institutions and individuals outside of the United States, Hawaii, and Alaska, while adequate for certain subjects, as a whole was extremely scattered and for this reason has been omitted (except as regards a few Canadian and Latin American members of the Association of American Geographers).

The *professional geographers* listed are, with a few exceptions, members of either the Association of American Geographers, the American Society for Professional Geographers, or both. These two organizations comprise the majority of the professional geographers of the country. Although many nongeographers were named in the answers, it was decided to omit them because of the difficulty of establishing any consistent principle as to whom to include. Obviously, however, certain nongeographers are as well versed as geographers in some of the subjects (for example, cartography, land utilization) under which the names of the latter appear.

The names assigned to any subject are those of (a) all persons who, in their own answers, specified that subject as one in which they have conducted research, (b) other American geographers whose familiarity with the subject is indicated in the answers, and (c) a few others added by the compiler. In most cases where a name is followed by a specific subject in parentheses, the person named listed the subject as one of his own special fields.

The purpose of the list is to present the information gathered through the questionnaire in an objective manner. Hence, except for additions, editorial changes have been kept to a minimum. In so far as possible the data are classified according to subjects *as indicated in the answers*. Since the names of many besides the respondents to the questionnaire are included, it is believed that the list provides a fairly satisfactory bird's-eye view of the various fields that American geographers are cultivating and

---

* This office has been discontinued; its geographical research functions have been taken over by the Department of State.

of the relative number of professional geographers in each. A few names, however, that are fully as worthy of inclusion under certain headings as are some of those actually listed have doubtless been overlooked.

Institutions mentioned in one or more answers as actively engaged in geographical research are designated by (g). The presence or absence of a (g), however, may well be misleading with regard to many institutions, as "geographical research" was not defined in the directions for filling out the questionnaire and was variously interpreted. Furthermore, since certain government agencies have been reorganized or discontinued and many geographers have returned from government service to the universities since the inquiry was made, the information regarding research is out-of-date in certain points of detail. Universities, government agencies, societies, and the like, mentioned as possessing libraries well equipped for purposes of geographical research are designated by (l) (or (gl) if they are also mentioned as engaged in research). L. means Library; P.L., Public Library; L.C., Library of Congress; U., University; and A.G.S., American Geographical Society. Other abbreviations are self-explanatory. A few local historical societies and libraries mentioned in the answers have been omitted.

Grateful acknowledgment is made to the officers of the Association of American Geographers and of the American Society for Professional Geographers for their collaboration in sending out the questionnaire, and especially to Professor R. H. Brown for his kindness in personally addressing the envelopes for members of the A.A.G. Sincere thanks are also due to all of those who answered the questionnaire. Many of the answers were given in detail at the cost of time and thought, and some were accompanied by friendly letters providing additional information. The following replied with much fullness and care: H. H. Bennett (on soil geography), D. D. Brand (on historical geography, agricultural history, and biogeography particularly concerning the New World and Mexico), J. W. Coulter (on work of the Committee on Pacific Science Appraisal of the National Research Council), S. de R. Diettrich (on the South, U.S.A.), C. D. Harris (on urban geography, the geography of manufacturing industries, the U.S.S.R., the Inter-mountain Region of the Western U.S.A., etc.), Owen Lattimore (on China), H. H. Rasche (on bibliographies of the Joint Intelligence Study Publishing Board), K. H. Stone (on aerial photo-interpretation and Alaska), and R. G. Stone (on climatology and bioclimatology). It is regretted that considerations of space and consistency made it necessary to omit some of the helpful information that these and others so kindly submitted.

GEOGRAPHICAL TOPICS

GENERAL TOPICS

History of Geography

Vilhjalmur Stefansson (geog. discovery, pre-Columbian relations between Europe and America). *Medieval geography:* G. H. T. Kimble, J. K. Wright. *Modern geography:* R. H. Brown, Richard Hartshorne, John Leighly. *History of cartography:* Clara E. Le Gear, F. J. Marschner, Lawrence Martin
A.G.S. (gl), Harvard U. (l), Huntington L., John Carter Brown L., L.C., Natn. Archives (l), New York P.L., U. Chicago (gl), U. Wisconsin (gl), Wm. L. Clements L. Various historical societies

Historical Geography

D. D. Brand (land settlement or occupance sequence), R. H. Brown (North America), G. S. Corfield, S. D. Dodge, C. O. Sauer
For institutions see History of Geography, above. For North American archeology: Bur. Amer. Ethnology (l), Peabody Mus. (Harvard U.) (l), Southwestern Mus. (Los Angeles) (l)

Methodology and Bibliography

(This list could be greatly expanded, since all geographers in one way or another give attention to these subjects.)

Isaiah Bowman, V. C. Finch, Richard Hartshorne, W. L. G. Joerg, John Leighly, R. S. Platt, G. T. Renner, Jr., J. K. Rose (quantitative techniques), C. O. Sauer, Derwent Whittlesey. *Bibliography:* Nordis Felland, J. K. Wright

Geographical Education

Mamie L. Anderzhon, W. W. Atwood, H. H. Barrows, N. A. Bengston, Benoît Brouillette, G. S. Corfield, R. E. Dodge, Alice Foster, Agnes Garrels (esp. elementary), G. D. Hubbard, G. J. Miller, Edith P. Parker, R. S. Platt, H. J. Warman, J. R. Whitaker
Clark U. (gl), Geo. Peabody Coll. (gl), Teachers Coll. (Columbia U.) (gl), U. Chicago (gl)

Cartography and Maps

S. W. Boggs, W. A. Briesemeister, G. B. Cressey, V. C. Finch, Eric Fischer, O. E. Guthe, R. E. Harrison (graphic methods, perspective cartography, nomographs), C. B. Hitchcock, J. R. Illick (standardization of geog. symbols), H. M. Leppard, R. S. McClure, Margaret Mace, F. J. Marschner, O. M. Miller, J. A. Minogue, C. B. Odell (representation of distribution of population), R. R. Platt, Erwin Raisz, A. H. Robinson, J. A. Russell, L. S. Silverman (geodetic control), G.-H. Smith, P. A. Smith, S. F. Smith, F. A. Stilgenbauer, H. T. Straw, R. J. Voskuil, L. S. Wilson, J. K. Wright. *Map projections:* S. W. Boggs, O. M. Miller. *Map information, cartobibliography:* S. W. Boggs, E. B. Espenshade, Jr., F. W. Foster, H. R. Friis, A. C. Gerlach, Anne M. Goebel, W. L. G. Joerg, Lawrence Martin, J. A. Minogue (official American cartography, Federal), W. W. Ristow, L. S. Wilson, Ena L. Yonge. (See also "History of Cartography.") *Map collections, classification and care of maps:* B. W. Adkinson, S. W. Boggs, E. B. Espenshade, Jr., Clara E. Le Gear, W. W. Ristow, A. H. Robinson, Ena L. Yonge. *Aerial photo-interpretation:* F. W. Foster, K. C. McMurray, J. A. Russell, K. H. Stone Amer. Automobile Assn. (1), A.G.S. (gl), Babson Inst. (g), Detroit P.L., L.C., Natn. Geog. Soc. (gl), Newberry L., New York P.L., Pan Amer. Union (1), Wisconsin Hist. Soc. (1). *Government agencies:* Army Air Forces (1), Army Engineers (1), Army Map Serv. (gl), Army War Coll. (1), Coast and Geodetic Surv. (g), Dept. of Agric. (gl), Dept. of Commerce (1), Dept. of State (gl), Geological Surv. (gl), Hydrographic Off. (Navy Dept.) (gl), Military Intell. Div. (1), Natn. Archives (1), *Universities:* California (1), Chicago (gl), Clark (1), Harvard (gl), Michigan (1), Ohio State (1), Yale (1), Wayne (1)

Geographical Names

Allen Belden, K. J. Bertrand, M. F. Burrill, Wilma B. Fairchild, W. L. G. Joerg, H. F. Raup, G. E. Reckford
U.S. Bd. on Geog. Names (gl)

PHYSICAL GEOGRAPHY AND BIOGEOGRAPHY

Physiography and Geomorphology

R. Van V. Anderson, W. W. Atwood, W. W. Atwood, Jr., G. B. Barbour, Isaiah Bowman, W. B. Brierly, Kirk Bryan (arid regions), W. S. Cole, A. B. Cozzens, G. B. Cressey (geog. aspects of land forms), R. M. Glendinning, C. B. Hitchcock, G. D. Hubbard (experimental physiography, shorelines, rivers), David Kai-Foo Loa, H. M. Kendall, J. E. Kesseli, John Leighly, A. K. Lobeck, K. F. Mather, F. E. Matthes, W. C. Mendenhall, H. A. Meyerhoff, J. J. Petty, Erwin Raisz, J. L. Rich, R. J. Russell, C. F. S. Sharpe (mass movement of soil and rock), G.-H. Smith, Anastasia Van Burkalow, O. D. Von Engeln (regional), Samuel Weidman, Bailey Willis, L. A. Wolfanger. *Glaciers and glacial land forms; Pleistocene and recent:* Ernst Antevs, W. J. Berry, Kirk Bryan (Pleistocene chronology), D. H. Davis, Hellmut de Terra, W. O. Field, Jr., Eric Fischer, G. D. Hubbard, K. F. Mather, H. A. Meyerhoff,

R. J. Russell, O. D. Von Engeln. *Streams and stream erosion:* R. M. Brown, G. D. Hubbard, Mark Jefferson (meandering streams), John Leighly, R. J. Russell
A.G.S. (1), Amer. Mus. Nat. Hist. (1), L.C., Smithsonian Instn. (g). *Government agencies:* Geological Surv. (gl), National Planning Assn. (g), Public Roads Admin. (1). *Universities, etc.:* California (gl), Chicago (gl), Cincinnati (gl), Clark (gl), Columbia (gl), Cornell (gl), Harvard (gl), Iowa (gl), Louisiana State (gl), Michigan (gl), Minnesota (g), Oberlin (g), Ohio State (1), Wisconsin (1), Yale (gl)

Oceanography

C. F. Brooks, P. E. Church, H. A. Marmer (currents, tides, mean sea level), P. A. Smith (ocean depths), J. C. Weaver (ice)
Harvard U. (gl), Scripps Inst. of Oceanography (gl), U.S. Coast and Geodetic Surv. (gl), U.S. Hydrographic Off. (gl), U. Washington (g), Woods Hole Oceanographic Instn. (gl)

Climatology and Meteorology

H. P. Bailey (machine tabulation techniques), M. H. Bissell, W. B. Brierly, C. F. Brooks, P. E. Church (snow cover), A. C. Gerlach, H. B. Hawkes (microclimatology, local winds), Ellsworth Huntington, David Kai-Foo Loa, H. M. Kendall, C. E. Koeppe, John Leighly, Hoyt Lemons, Peveril Meigs, R. W. Richardson, J. K. Rose (weather and crop yields), R. J. Russell, R. G. Stone (tropical climatology, applied bioclimatology, snow cover), Griffith Taylor, C. W. Thornthwaite, G. T. Trewartha, Eugene Van Cleef, Samuel Van Valkenburg, S. S. Visher (cl. and its influences: climatic changes, tropical cyclones, precip. intensity, regional climates, cl. of Indiana), H. C. Willett
A.G.S. (1), Amer. Geophys. Union (g), Blue Hill Observatory (gl), L.C., New York P.L., Woods Hole Oceanographic Instn. (g). *Government agencies:* Aero-Medical Lab. (Wright Field) (g), Army Air Forces (g), Bur. of Plant Ind. (g), Chemical Warfare Serv. (g), Climatic Research Div. (Q.M. Gen.) (g), Dept. of Agric. (gl), Forest Serv. (g), Pub. Health Serv. (g), Signal Corps (g), Soil Conservation Serv. (g), Weather Bur. (gl), Weather Research Center (War Dept.) (g). *Universities:* California (g), California (Los Angeles) (g), California Inst. of Tech. (gl), Chicago (gl), Clark (1), Harvard (gl), Massachusetts Inst. of Tech. (g), Michigan (1), New York U. (g), Rutgers (g), Yale (gl)

Plant Geography

W. B. Brierly, S. A. Cain, W. S. Cooper, W. E. Ekblaw, M. L. Fernald, H. M. Raup, H. L. Shantz, V. E. Shelford, Forrest Shreve, E. N. Transeau, S. S. Visher (biogeography). *Origins of American agriculture:*

D. D. Brand, G. F. Carter, C. O. Sauer. *Deserts and steppes:* Ellsworth Huntington, Griffith Taylor
A.G.S. (1), Desert Bot. Garden (Carnegie Instn., Wash.) (g), L.C., Missouri Bot. Garden (g), New York Bot. Garden (gl), New York P.L., U.S. Dept. of Agric. (1). *Universities:* California (gl), Cincinnati (g), Clark (1), Harvard (gl), Illinois (gl), New Mexico (g), Wisconsin (gl), Yale (gl)

Animal Geography

R. C. Murphy, H. L. Shantz (big-game population)
Amer. Mus. of Nat. Hist. (gl)

HUMAN GEOGRAPHY

Medical Geography

(This field has been developed primarily by medical scientists and other nongeographers among whom may be listed G. W. Anderson (U. Minn.), Zygmunt Deutschman, Saul Jarcho. The American Geographical Society is inaugurating studies of the geography of diseases.)

W. B. Brierly, Ellsworth Huntington
Harvard U. (1), Johns Hopkins U. (Sch. of Pub. Health) (g), L.C., Natn. Inst. of Health (g), New York Acad. of Medicine (1), U.S. Pub. Health Serv. (Surgeon General's L.)

Population and Settlements

Population: C. E. Batschelet, Allen Belden, W. J. Berry (world distribution of pop.), J. F. Bogardus, L. A. Hoffman (world population: total; changes; by occupations), Mark Jefferson (growth and distribution of pop.), C. B. Odell (cartographic representation). *Races, religions, languages, etc.:* Ellsworth Huntington, Griffith Taylor
A.G.S. (1), L.C., League of Nations Off. (Princeton, N.J.) (1), Miami U. (g), Millbank Fdn. (g), Off. of Population Research (Princeton, N.J.) (gl), Princeton U. (School of Public Affairs) (gl), Scripps Fdn. (g), U.S. Bur. of Census (gl), U.S. Dept. of State (gl)
Settlement: D. D. Brand (historical), S. D. Dodge, F. B. Kniffen, C. F. Kohn, W. M. Kollmorgen, G. T. Trewartha. *Migration and pioneering:* Isaiah Bowman, K. J. Pelzer, Griffith Taylor
A.G.S. (1), Harvard U. (1), Johns Hopkins U. (1), New York P.L., U. of Wisconsin (gl)

Urban Geography

J. Q. Adams, T. F. Barton, R. M. Brown, Edna F. Campbell, W. T. Chambers, C. C. Colby, W. E. Ekblaw (water supply), C. D. Harris, Bert Hudgins (city water supplies), Mark Jefferson, L. E. Klimm, R. C. Klove (urban and rural urban geog.), E. W. Miller, R. E. Murphy,

L. O. Myers, Rafael Picó, M. J. Proudfoot, Victor Roterus, F. A. Stilgenbauer, H. T. Straw, Griffith Taylor, L. F. Thomas, Edward Ullman, Eugene Van Cleef, C. M. Zierer
A.G.S. (1), Amer. Soc. of Planning Officials (gl), Carnegie L. of Pittsburgh, Chicago Planning Commission (g), Cleveland P.L., Detroit P.L., John Crerar L., L.C., Newberry L., Bur. of Urban Research (Princeton U.) (g), Pub. Admin. Clearing House (Chicago) (g), U.S. Bur. of Census (1), U.S. Natn. Housing Agency (g). *Universities:* Chicago (gl), Harvard (1), Ohio State (gl), Washington (St. Louis) (1), Wayne (gl), Western Reserve (g)

Economic Geography

A. K. Botts, A. S. Carlson, E. C. Case, C. C. Colby, G. S. Corfield, S. de R. Diettrich, S. T. Emory, V. C. Finch, Merna I. Fletcher, N. A. Bengtson, R. B. Hall, Margaret A. Hitch, G. D. Hudson, Ellsworth Huntington, L. E. Klimm, R. W. Johnson, C. F. Jones, H. H. McCarty, O. M. McMillion (foods), P. C. Morrison, J. E. Orchard, H. F. Otte, Rafael Picó, G. T. Renner, Jr., Victor Roterus, V. E. Shelford, G.-H. Smith, J. R. Smith, F. A. Stilgenbauer, L. F. Thomas (commercial geog.), E. N. Torbert, L. W. Trueblood, Eugene Van Cleef, S. S. Visher, J. R. Whitaker, F. E. Williams, A. J. Wright, E. W. Zimmerman (world resources and industries). *Tropics:* K. J. Pelzer, Leo Waibel
A.G.S. (1), Carnegie L. of Pittsburgh, Cleveland P.L., Detroit P.L., L.C., Minneapolis P.L. *Government agencies:* Bur. of Agric. Econ. (gl), Bur. of Labor Statistics (g), Dept. of Commerce (1), Dept. of Interior (1), Off. of Foreign Agric. Relations (1). *Universities:* California (1), Chicago (gl), Cincinnati (1), Clark (gl), Dartmouth (gl), Florida (gl), Harvard (gl), Maryland (g), Michigan (gl), Minnesota (1), Nebraska (gl), Ohio State (gl), Pennsylvania (gl), Stanford (gl), Washington (St. Louis) (gl), Wayne (1), Yale (1)

Land Utilization

O. E. Baker, C. P. Barnes, M. F. Burrill, C. C. Colby, J. W. Coulter, W. E. Ekblaw, V. C. Finch, E. J. Foscue, J. S. Gibson, H. A. Hoffmeister (arid lands), R. W. Johnson, W. D. Jones, K. C. McMurry, J. A. Minogue, R. S. Platt, H. L. Shantz. *Tropics:* R. G. Bowman, K. J. Pelzer, R. L. Pendleton, C. O. Sauer, Leo Waibel
Clark U. (gl), L.C., Natn. Archives, U.S. Dept. of Agric. (gl), U. California (gl), U. Chicago (gl), U. Michigan (gl), U. Wisconsin (gl)

Land and Regional Planning (Data furnished by C. C. Colby)

C. P. Barnes, H. H. Bennett, M. F. Burrill, C. C. Colby, Loyal Durand, Jr., W. E. Ekblaw, R. M. Glendinning, G. D. Hudson, K. C. McMurry,

F. J. Marschner, M. J. Proudfoot, Victor Roterus, E. N. Torbert, J. O. Veatch, J. R. Whitaker

Natn. Planning Assn. (g), Pub. Admin. Clearing House (Chicago, New York) (l), U. Chicago (l). *Government agencies:* Bur. of Agric. Econ. (g), Bur. of Reclamation (g), Forest Serv. (g), Genl. Land Office (g), Geological Surv. (g), Soil Conserv. Serv. (g), Soil Surv. (g)

Natural Resources: Conservation, Development, and Management

E. A. Ackerman, J. F. Bogardus, N. B. Guyol (energy resources of world), J. K. Rose, J. R. Whitaker (conservation history and theory). *Water supplies:* C. Y. Mason, Anastasia Van Burkalow, G. F. White

A.G.S. (l), Harvard U. (l), L.C., U. Wisconsin (l)

Geography of Mineral Resources

C. H. Behre, Jr., J. F. Bogardus, J. E. Collier, J. W. Frey (petroleum), P. W. Icke, E. W. Miller (petroleum), R. E. Murphy, H. E. Spittal (coal), Warren Strain, H. W. Straley, W. H. Voskuil

Brookings Inst. (g), Carnegie L. of Pittsburgh, Cleveland P.L., Columbus P.L., Engineering Societies (New York) (l), Illinois State Geol. Soc. (g), John Crerar L., L.C., Michigan Sch. of Mines (l). *Government agencies:* Bur. of Mineral Industries (Dept. of Interior) (g), Bur. of Mines (gl), Dept. of State (g), For. Econ. Admin. (g), Geological Surv. (gl), War Production Bd. (g). *Universities, etc.:* Chicago (l), Maryland (g), Missouri (gl), Pennsylvania State (g).

Geography of Agriculture and Forests (See also Land Utilization)

O. E. Baker, M. K. Bennett, Isaiah Bowman, R. H. Brown, Elizabeth Day, S. N. Dicken, V. C. Finch, W. D. Jones (classification of world agric. regions), Richard Hartshorne, W. M. Kollmorgen, Peveril Meigs, Lois Olson, R. S. Platt, G. H. Primmer, C. S. Scofield (irrigation), J. R. Smith, Warren Strain, Leo Waibel, J. C. Weaver, Derwent Whittlesey. *Historical geography of agriculture:* D. D. Brand, C. O. Sauer

A.G.S. (l), L.C., New York P.L. *Government agencies:* Dept. of Agric. (gl), Forest Serv. (g), various agric. experiment stations (gl). *Universities, etc.:* Chicago (gl), Columbia (l), Cornell (g), Harvard (g), Illinois (gl), Michigan Agric. (l), Missouri (l), Pennsylvania (l), Stanford (g), Wisconsin (gl); various agric. colleges (gl)

Soil Geography and Soil Conservation

C. P. Barnes, H. H. Bennett, W. B. Brierly, W. E. Ekblaw, W. D. Jones, C. E. Kellogg, K. C. McMurry, Konstantin Nikiforoff, Lois Olson (history of soils), James Thorp (paleopedology), J. O. Veatch. *Soil conservation:* H. H. Bennett, C. W. Thornthwaite, S. S. Visher

L.C., U.S. Bur. of Reclamation (1), U.S. Dept. of Agric. (gl), U.S. Soil
Conservation Serv. (g), U.S. Soil Surv. (g). *Universities, etc.:* California
(gl), Chicago (gl), Cornell (gl), Illinois (gl), Iowa State Coll. (gl),
Michigan State Coll. (gl), Ohio State (gl), Rutgers (gl), Wisconsin
(gl)

Geography of Fisheries

E. A. Ackerman, S. W. Boggs, Ada V. Espenshade, G. W. Hewes
Harvard U. (1), U.S. Bur. of Fisheries (gl), U.S. Fish and Wildlife Serv.
(gl), U. Washington (l)

Industrial Geography: Geography of Manufacturing Industries

M. F. Burrill, Loyal Durand, Jr., N. B. Guyol (energy and industrial re-
sources of the world), C. D. Harris, Richard Hartshorne, J. K. Rose,
J. A. Russell, J. R. Smith, F. A. Stilgenbauer, Helen M. Strong, L. F.
Thomas, C. L. White, A. J. Wright, C. M. Zierer, E. W. Zimmerman
A.G.S. (1), Detroit P.L., U.S. Army Industrial Coll. (g), U.S. Bur. of
For. and Domestic Com. (g), U.S. Dept. of Commerce (1), Wayne
U. (1)

Geography of Transportation and Trade

R. M. Brown C. C. Colby (ocean trade and transportation), C. B. Odell
(commodity distribution), R. S. Platt, Edward Ullman

Political Geography

S. W. Boggs, Isaiah Bowman, J. O. M. Broek, A. S. Carlson, W. T. Chambers (geopolitics), G. B. Cressey, S. de R. Diettrich, Richard Hartshorne, G. D. Hubbard (geog. aspects of world peace), S. B. Jones,
Geoge Kiss, R. S. Platt, Evelyn L. Pruitt, Marie Santes, Sophia Saucer-
man, G.-H. Smith, L. W. Trueblood, Edward Ullman, Samuel Van
Valkenburg, H. W. Weigert, Derwent Whittlesey (polit. geog. inc.
hist. aspects), J. K. Wright. *International boundaries:* S. W. Boggs,
Isaiah Bowman, Richard Hartshorne, S. B. Jones
A.G.S. (gl), Council on For. Relations (g), For. Policy Assn. (g), Inst.
of Pac. Relations (g), L.C., U.S. Dept. of State (l). *Universities, etc.:*
Chicago (1), Clark (1), Columbia (1), Florida (gl), Harvard (gl),
Michigan (1), Missouri (1), Oberlin (1), Ohio State (1), Stanford
(Hoover L.), Wisconsin (g), Yale (g).

Military Geography

The following names of geographers who, besides himself and Col.
S. P. Poole, are or have been employed upon studies of military geography in the Military Intelligence Section, War Dept., were kindly
furnished by G. F. Deasy

J. Q. Adams, Margaret Hitch, C. Y. Hu, H. M. Kendall, John Kesseli, L. E. Klimm, F. W. McBryde, L. H. Parkhurst, J. A. Russell, O. P. Starkey, L. W. Trueblood, Joseph Van Riper, W. A. Wallace

REGIONS

Polar Regions

W. L. G. Joerg, Vilhjalmur Stefansson. *Arctic:* W. E. Ekblaw, L. M. Gould (Baffin I., Greenland), W. H. Hobbs (Greenland), Trevor Lloyd (Canadian Arctic, Greenland), A. L. Washburn. *Antarctic:* L. M. Gould, W. H. Hobbs, Lawrence Martin (history of exploration), R. C. Murphy, P. A. Siple, Griffith Taylor
A.G.S. (1), Arctic Inst. of North America (g), Carleton Coll. (1), Explorers' Club (New York) (1), John Crerar L., L.C., V. Stefansson's personal L. (New York), U. Michigan (1)

Pacific Region (See also Asia, Far East, Australia and New Zealand)

J. W. Coulter (islands), O. W. Freeman (political), H. E. Gregory (islands), W. H. Hobbs (western Pac.), S. B. Jones (Hawaii and S. W. Pac.), C. R. MacFadden (maps), R. C. Murphy (zoogeog.), Griffith Taylor, Samuel Van Valkenburg, S. S. Visher (west coast and islands), C. M. Zierer
A.G.S. (1), Amer. Mus. Nat. Hist. (g), Bishop Mus. (gl), Inst. of Pacific Relations (g), John Crerar L., L.C., Smithsonian Instn. (gl). *Universities:* California (gl), California (Los Angeles) (g), Hawaii (gl), Stanford (gl), Washington (gl)

North America

Under the headings "North America" and "United States" there are included (with a few exceptions) only the names of persons who, in answering the questionnaire, specified those areas among their special fields. For the names of other individuals and of libraries and research institutions pertinent to North America and the United States the topical section should be consulted.

O. E. Baker (population), H. H. Bennett (soils), W. J. Berry, C. C. Colby (economic geog.), E. J. Foscue, H. B. Hawkes (western N. Amer. Utah), W. L. G. Joerg, C. F. Kohn, O. M. McMillion, L. O. Myers, C. B. Odell, Victor Roterus, C. O. Sauer, J. R. Smith, L. F. Thomas, S. S. Visher (geog. of Amer. notables), Clark Wissler (culture areas), A. J. Wright. *Agricultural geography and land use:* O. E. Baker, W. A. Rockie (western North America and Mexico, soil conservation and plant ecology). *Climates:* C. F. Brooks, C. W. Thornthwaite. *Geomorphology:* W. W. Atwood, W. S. Cooper (Pacific coast, dunes), W. H. Hobbs (glaciated areas), T. W. Vaughan (Atlantic and Gulf Coastal Plain). *Historical geography:* R. H. Brown, G. W. Hewes (archeol.). *Plant geography:* C. P. Barnes (native vegetation), H. A. Gleason (eastern N. Amer.), V. E. Shelford

Alaska

M. F. Burrill, W. A. Rockie (soil conservation, plant ecology), R. H. Sargent, P. S. Smith (geology, physiography), K. H. Stone. *Southeast Alaska: glaciers, vegetation:* W. S. Cooper, W. O. Field, Jr. A.G.S. (1), Carnegie Instn. of Washington (g), Harvard U. (1), L.C., Smithsonian Instn. (g), Territorial L. (Juneau) (1), U.S. Geological Surv. (gl), U. Alaska (g)

Canada

Benoît Brouillette (economic and regional geog.; Prov. of Quebec), A. H. Clark (Prince Edward I., hist. geog.; Laurentian upland, geomorphology), Bert Hudgins (St. Clair delta and vicinity), O. E. Jennings (N. shore, L. Superior), G. H. T. Kimble, Trevor Lloyd, D. F. Putnam (Maritime Provinces), Griffith Taylor, J. R. Whitaker (central Can.). *French Canada:* C. B. Odell, Roderick Peattie. *Plant geography:* M. L. Fernald, H. M. Raup

Boston P.L., Harvard U. (gl), L.C., New York Bot. Garden (plant geog.) (gl), New York P.L., Wayne U. (1), Wisconsin Hist. Soc. (1), Yale U. (1)

United States (See also North America)

W. A. Browne (historical geog.), F. B. Kniffen (material forms of landscape), J. L. Rich (eastern U.S.), Warren Strain, S. S. Visher, C. L. White. *Agricultural geography:* C. P. Barnes, Loyal Durand, Jr. (dairy regions). *Economic geography:* J. B. Appleton (industrial geog., regional planning), G. D. Hubbard (precious metals), H. H. McCarty (econ. regions), C. W. Miller (eastern U.S.). *Land utilization:* J. D. Abrahamson, J. H. Garland, K. C. McMurry, F. J. Marschner (physical conditions and land use), J. A. Minogue. *Soils:* J. K. Ableiter (soil classification), C. P. Barnes, C. E. Kellogg, James Thorp

The West

Ruth E. Baugh (urban geog.), W. T. Buckley, G. F. Carter (southwest), E. J. Foscue, Konstantin Nikiforoff, C. O. Sauer, J. H. Steward (anthropology), O. C. Stewart (cultural geog.), Bailey Willis (geomorphology)

Knox Coll. (1), San Diego Mus. (1), Stanford U. (1), U. California (1)

Pacific Northwest

J. B. Appleton, J. H. Bretz (geomorphology), J. P. Carey, O. W. Freeman (mineral resources, physiography, econ. geog.), J. H. Garland, C. H. Mapes, H. F. Raup (econ. geog), W. A. Rockie, V. E. Shelford (Puget Sound)

L.C., Northwest Regional Council (g), Northwest Scientific Assn. (g), Pacific Northwest Forest and Range Exper. Sta. (g), Pacific Northwest Regional Planning Bd. (g). *Universities, etc.:* Chicago (gl), E. Washington Coll. of Educ. (1), Idaho (gl), Oregon (1), Oregon State Coll. (1), Washington (1), Washington State Coll. (1)

Intermountain Region

Ernst Antevs, M. H. Bissell, Eliot Blackwelder (southwestern desert region), G. F. Carter (Mohave and Colorado desert areas), W. E. Coffman, H. E. Gregory (Colorado Plateau), C. D. Harris, J. W. Hoover, C. L. White

Brigham Young U. (1), Bur. of Econ. and Bus. Research (U. Utah) (g), Carnegie Instn. of Washington (g), Huntington L., L.C., L. of Church Historian's Off. (Church of J.C. of L.D.S., Salt Lake City) (1), San Diego Mus. (for Mohave and Colorado desert areas) (1), Smithsonian Instn. (1), U. Utah (gl), Utah State Agric. L., Utah State Planning Bd. (g)

The Rocky Mountains

C. M. Davis (southern Rocky Mountains), H. A. Hoffmeister. *Geomorphology and geology* (see also N. Amer.): W. W. Atwood, W. W. Atwood, Jr., C. H. Behre, Jr., Eliot Blackwelder, Kirk Bryan, L. O. Quam, H. S. Sharp

Colorado Hist. Soc. (1), Denver P.L. *Universities, etc.:* Chicago (g), Clark (gl), Colorado (1), Colorado Agric. Coll. (g), Columbia (gl), Harvard (g), Princeton (g), Wyoming (1)

North Central United States

Esther S. Anderson (econ. geog), T. F. Barton, C. B. Odell (corn belt), L. F. Thomas, W. H. Voskuil. *Pleistocene:* W. S. Cooper, L. M. Gould

Illinois State Geol. Surv. (g), Missouri Hist. Soc. (1), St. Louis P.L. *Universities, etc.:* Carleton Coll. (1), Minnesota (1), Washington U. (1)

Great Plains Region

R. H. Brown, H. A. Hoffmeister, H. L. Shantz, C. W. Thornthwaite, C. L. White

Colorado Hist. Soc. (1), Minnesota Hist. Soc. (1), U. Wisconsin (1)

Great Lakes Region

W. O. Blanchard, O. E. Guthe, J. A. Minogue (land utilization, cut-over lands), P. C. Morrison, G. H. Primmer (upper Lakes reg.), H. H. Rasche (urban geog.), J. O. Veatch

Minnesota Hist. Soc. (l), U. Chicago (l), U. Wisconsin (gl), Wisconsin Hist. Soc. (l)

The South

R. S. Atwood, W. A. Browne (econ. geog.), S. de R. Diettrich (esp. Florida), J. S. Gibson, A. R. Hall (hist. geog.), C. F. Kohn, W. M. Kollmorgen (agric. geog.), B. F. Lemert, M. C. Pronty, Jr., R. J. Russell (lower Mississippi Valley), Irma Scott, H. T. Straw, J. A. Tower, Edward Ullman, J. R. Whitaker, L. S. Wilson. *Appalachian region:* R. E. Murphy (regional geog.), H. S. Sharp (geomorphology), C. F. S. Sharpe (Atlantic Piedmont, soil conservation), F. J. Wright (southern Appals., geomorphology)

Alabama State Archives (l), Georgia Land Office (l), Historical commissions of Georgia, North Carolina, and Virginia (l), L.C. *Universities:* Alabama (Bur. of Bus. Research) (g), Chicago (gl), Duke (l), Florida (gl), Harvard (l), Michigan (l), North Carolina (gl), Vanderbilt (l), Virginia (l), Wisconsin (l)

Tennessee Valley: H. C. Amick (geology and econ. geog.), G. D. Hudson, H. V. Miller

George Peabody Coll. (g), Joint Reference L. (Nashville), Tennessee Planning Commission L. (Nashville) (gl), T.V.A. Reference L. (Knoxville), U. Tennessee (gl)

New England

J. H. Burgy (industrial geog.), J. P. Carey (upper Connecticut Valley), A. S. Carlson, S. D. Dodge, W. E. Ekblaw, M. L. Fernald (plant geog.), J. W. Goldthwait (geomorphology), J. R. Illick (forest industries), H. S. Kemp, C. F. Kohn, J. A. Minogue (land utilization), H. M. Raup (forest history), Derwent Whittlesey

A.G.S. (l), New England Planning Assn. (g). *Universities, etc.:* Bowdoin (l), Clark (g), Columbia (l), Dartmouth (gl), Harvard (gl), Maine (g), New Hampshire (g), Yale (gl)

Individual States

Arizona: Agnes M. Allen (human geog.), F. B. Kniffen

Arizona State Teachers Coll. (l), Northern Arizona L. (g), Northern Arizona Soc. of Sci. and Art (l)

California: R. Van V. Anderson, J. O. M. Broek, J. E. Kesseli, F. B. Kniffen, Peveril Meigs, W. H. Miller, Evelyn L. Pruitt, H. F. Raup (econ. and hist. geog.), Forrest Shreve (plant geog.), C. M. Zierer (southern Calif.). *Geomorphology:* Eliot Blackwelder (Sierra Nevada), F. E. Matthes

California Acad. of Sci. (l), California Hist. Soc. (l), Carnegie Instn. of Washington (g), Huntington L. (gl), Stanford U. (gl), U. California (gl)

District of Columbia: J. A. Minogue (urban geog. of Washington)
Columbia Hist. Soc. (1), L.C., Natn. Archives (1), Natn. Parks and Planning Com. (g), Washington P.L.
Florida: S. de R. Diettrich, L. S. Silverman
Idaho: W. M. Kollmorgen (southern Idaho, agric.)
Illinois: Edna F. Campbell (Chicago: econ. and hist. geog.), Alden Cutshall, G. T. Trewartha
John Crerar L., Newberry L., U. Chicago (gl), U. Illinois (gl)
Indiana: Alden Cutshall, G. D. Koch (meteorology), S. S. Visher
Indiana State L., Indiana U. (gl)
Iowa: H. H. McCarty (econ. geog.)
U. Iowa (gl)
Kentucky: S. N. Dicken
Filson Club (Louisville) (gl), U. Kentucky (l)
Louisiana: Edna F. Campbell (New Orleans: econ. and hist. geog.), F. B. Kniffen, R. J. Russell
Louisiana State U. (gl), Tulane U. (l)
Maryland: H. J. Warman (manor counties)
Clark U. (1), Johns Hopkins U. (l)
Michigan: J. P. Carey, Bert Hudgins, K. C. McMurry, J. O. Veatch
Detroit P.L., Michigan Agric. Exper. Sta. (g), Michigan Conserv. Dept. (g), Michigan Sch. of Mines (g), Michigan State Coll. (l), U. Michigan (l)
Minnesota: D. H. Davis
Minnesota State Hist. Soc. (1), U. Minnesota (gl)
Mississippi: J. A. Minogue (land utilization)
Nebraska: N. A. Bengtson, G. D. Koch (meteorology, loess hill country), E. E. Lackey, W. A. Rockie (soil geog.)
U. Nebraska (gl)
New York: O. D. Von Engeln (regional geog. central New York). *Western New York:* M. Melvina Svec, K. T. Whittemore
Buffalo Hist. Soc. (1), Buffalo P.L. (western N.Y.), U. Buffalo (l)
Ohio: Mildred Danklefsen (northern Ohio), R. B. Frost, J. H. Garland, C. C. Huntington, Shannon McCune, G.-H. Smith, Eugene Van Cleef, A. J. Wright
Ohio State L., Ohio State U. (gl), Western Reserve Hist. Soc. (1), Western Reserve U. (gl)
Oklahoma: Leslie Hewes (Cherokee country, cultural and hist. geog.)
Oklahoma State Hist. Soc. (1), U. Oklahoma (gl)
Pennsylvania: O. E. Jennings (ecological plant geog.), W. M. Kollmorgen (agric. geog.), L. O. Myers, R. E. Murphy, (regional geog.), H. F. Raup (hist. geog.), O. P. Starkey (eastern Pennsylvania), F. E. Williams
Carnegie Museum of Pittsburgh (g), Hist. Soc. of Pennsylvania (gl),

Pennsylvania State Coll. (gl), State L. (Harrisburg), U. Pennsylvania (1)
South Carolina: J. J. Petty (demography)
U. South Carolina (1)
South Dakota: S. S. Visher
Tennessee (see also Tennessee Valley): H. C. Amick, J. R. Whitaker, L. J. Zuber (place names)
Cossitt L. (Memphis), East Tennessee Hist. Soc. (g), Tennessee Hist. Soc. (g), Tennessee State L., U. Chattanooga (1), U. Tennessee (1)
Texas: C. J. Bollinger, W. T. Chambers, E. J. Foscue
Southern Methodist U. (1), Stephen F. Austin State Teachers' Coll. (1)
Utah: H. B. Hawkes, W. M. Kollmorgen (northern Utah, agric.), J. E. Spencer (southern Utah, cultural geog.)

CARIBBEAN AREA

H. H. Bennett (soil geog.), C. F. Brooks (climatology), F. A. Carlson, R. E. Crist, J. E. Fairchild (regional and econ. geog.), E. P. Hanson, Rafael Picó, William Van Royen, T. W. Vaughan, E. W. Zimmerman (Puerto Rico). *Cuba:* Salvador Massip, Mrs. Salvador Massip. *Lesser Antilles, economic geography:* O. P. Starkey, F. E. Williams
A.G.S. (gl), Clark U. (1), Columbia U. (gl), Inst. of Trop. Agric. (U. Puerto Rico) (g), Insular Experiment Sta. (San Juan, P.R.) (1), L.C., Middle Amer. Research Inst. (Tulane U.) (1), New York P.L., New York U., Pan Amer. Union (gl), Stanford U. (1), United Fruit Co. (g), U. Chicago (g), U. Pennsylvania (1), U. Puerto Rico (1). *Government agencies:* Dept. of Agric. (1), Dept. of Commerce (1), Div. of Territories and Island Possessions (1), Geological Survey (1), Soil Conserv. Serv. (g), Soil Surv. (g)

LATIN AMERICA

J. D. Abrahamson, R. S. Atwood, G. A. Beishlag, H. H. Bennett (soils), A. P. Biggs (tropical highlands), Isaiah Bowman, D. D. Brand, H. J. Bruman, F. A. Carlson, G. F. Carter (origins of American agric.), R. E. Crist, S. de R. Diettrich, Wilma B. Fairchild, E. J. Foscue (econ. and regional geog.), H. A. Gleason (plant geog., northern South Amer.), W. H. Haas, E. P. Hanson, P. E. James, C. F. Jones, G. M. McBride, F. W. McBryde, O. M. McMillion, K. F. Mather, R. L. Pendleton (Peru and Ecuador, soils and land use), J. J. Petty (demography), R. S. Platt, R. C. Murphy (zoogeog.), Rafael Picó, Evelyn L. Pruitt, J. L. Rich, R. W. Richardson, C. O. Sauer, E. B. Shaw, Dan Stanislawski (historical geog.), H. S. Sterling, J. H. Steward (anthropology), Leo Waibel, H. J. Warman, C. L. White. *Brazil:* R. E. Crist, C. A. Gauld (Minas Gerais and Brazilian Plateau, econ. and social geog.), P. E. James. *Cartography:* C. B. Hitchcock, R. R. Platt. *Central America:* E. J. Foscue, G. M. McBride, F. W. McBryde, R. L. Pendleton (soils and land use), Leo Waibel. *Mexico:* D. D. Brand, R. E. Crist, S. N. Dicken, Alice Foster, B. F. Lemert, G. M. McBride, W. A. Rockie (soil conservation and plant ecology), C. O. Sauer, V. E. Shelford (northern Mexico), Forrest Shreve (plant geog.), H. S. Sterling (Central Mexico, land tenure). *Northwestern Mexico:* Leslie Hewes, F. B. Kniffen, Peveril Meigs (Baja Calif.)

A.G.S. (gl), Carnegie Instn. of Washington (gl), Huntington L., John Carter Brown L., L.C. (Hispanic Foundation), Natl. Planning Bd. (g), New York P.L., Pan Amer. Inst. of Geog. and Hist. (gl), Pan Amer. Union (gl), Smithsonian Instn. (g), United Fruit Co. (Middle Amer. Research Inst., N.Y.C.), U.S. Bur. of For. and Dom. Com. (g), U.S. Dept. of Agric. (gl), U.S. Dept. of Com. (l), U.S Soil Conserv. Serv. (g). *Universities:* California (Bancroft L.) (gl), California (gl), California (Los Angeles) (l), Cincinnati (l), Chicago (gl), Clark (gl), Columbia (l), Florida (l), Harvard (gl), Johns Hopkins (l), Michigan (gl), Middle Amer. Research Inst. (Tulane U.) (gl), Pennsylvania (l), Puerto Rico (l), Texas (l), Western Reserve (l)

EUROPE

M. K. Bennett, W. O. Blanchard (econ. geog.), Eric Fischer (east-central and southeastern Eur.), J. H. Garland (British Isles), W. T. Gehrke (climate), Richard Hartshorne, G. D. Hubbard, George Kiss, Olga Kutby (Hungary), J. B. Leighly (northern Eur., urban geog.), G.-H. Smith, O. P. Starkey (econ. and physical geog.), M. Melvina Svec (Slavic Eur., human geog.), Eugene Van Cleef (Finland, Finns in U.S.), William Van Royen, Samuel Van Valkenburg, S. S. Visher, Derwent Whittlesey. *Western Europe:* J. B. Appleton (econ. and industrial geog.), J. F. Bogardus (econ. geog.), S. D. Dodge, Jean Gottmann (France), C. D. Harris, H. M. Kendall, L. E. Klimm (Ireland), H. M. Leppard, L. O. Myers, Roderick Peattie (mountains). *Alsace-Lorraine:* Marie Santes (bibliography). *Germany:* H. B. Hawkes (regional geog.), L. A. Hoffman (population, industrial geog.). *U.S.S.R.* (as a whole): G. B. Cressey, Merna I. Fletcher, C. D. Harris, L. A. Hoffman (population), J. A. Morrison

A.G.S. (gl), Brookings Instn. (g), John Crerar L., L.C. (Div. of Slavic Literature), New York P.L., Russian Econ. Inst. (g), U.S. Dept. of Agric. (gl), U.S. Dept. of State (gl), U.S. Office of Strategic Serv. (g). *Universities:* California (l), Chicago (gl), Clark (l), Columbia (l), Harvard (l), Illinois (gl), Michigan (gl), Missouri (l), Ohio State (l), Stanford (g), Syracuse (gl), Wisconsin (gl), Yale (l)

Appendix

Asia (See also Far East; for U.S.S.R., see Europe)

G. B. Barbour (physiography, China and adjacent regions), J. W. Coulter, G. B. Cressey, Hellmut de Terra (early migrations), R. B. Hall, Ellsworth Huntington, W. D. Jones, David Kai-Foo Loa, George Kiss (China, Japan, Middle East), J. E. Orchard, J. A. Russell (Japan, Japanese Empire, regional, urban), G. T. Trewartha, Samuel Van Valkenburg. *China:* G. B. Cressey, J. E. Fairchild, W. T. Gehrke (transportation), George Kiss, O. D. Lattimore, J. L. Rich (physiography), J. R. Smith, J. E. Spencer, James Thorp (soil geog.). *India:* R. Van V. Anderson, L. A. Hoffman (population, econ. geog.), W. D. Jones, Warren Strain. *Japan:* J. D. Abrahamson, Mamie L. Anderzhon, G. A. Beischlag, H. W. Burkland, D. H. Davis, Virginia M. Dewey, W. T. Gehrke (transportation), R. B. Hall, G. W. Hewes (place names), L. A. Hoffman (population), George Kiss, J. E. Orchard, J. A. Russell, G.-H. Smith, G. T. Trewartha. *Southeastern Asia:* J. O. M. Broek, Hellmut de Terra (physiography, geog. exploration), K. J. Pelzer, R. L. Pendleton (soils and land use), Helen L. Smith (place names), L. W. Trueblood. *Western Asia:* Hellmut de Terra, Jean Gottman (Palestine), J. A. Tower, Bailey Willis (geomorphology)

A.G.S. (l), Inst. of Pacific Relations (g), Iranian Inst. (New York) (gl), L.C., New York P.L., Office of Population Research (Princeton), Sch. of Asiatic Studies (New York) (g). *Universities:* California (l), California (Los Angeles) (l), Chicago (gl), Clark (l), Columbia (gl), Harvard (gl), Michigan (gl), Syracuse (l), Wisconsin (gl), Yale (l)

Far East (See also Pacific Region, Asia)

J. B. Appleton (econ. and polit. geog.), A. H. Clark (econ. geog.), Frances M. Earle, E. M. Harwood, Jr. (meteorology), G. W. Hewes (China, Japan, Korea, North Pacific, esp. fisheries). W. D. Jones, O. D. Lattimore (China and Japan), Shannon McCune, E. W. Miller (cartography), K. J. Pelzer (agric. geog.), R. L. Pendleton (soils geog.), J. E. Spencer (cultural and hist. geog.), G. T. Trewartha. *Netherlands Indies:* J. O. M. Broek, Elizabeth Day, K. J. Pelzer, Joseph Van Riper. *Philippine Islands:* Alden Cutshall (econ. geog.), K. J. Pelzer, Bailey Willis (geomorphology)

A.G.S. (gl), Div. of Orientalia (L.C.) (gl), Inst. of Pacific Relations (g), L.C., New York P.L., Southeast Asia Institute (g). *Universities, etc.:* California (gl), Chicago (gl), Claremont (l), Clark (l), Columbia (l), Harvard (gl), Illinois (l), Louisiana State (l), Michigan (gl), Stanford (l), Washington (gl), Wisconsin (gl), Yale (l)

AFRICA

H. H. Bennett (soils), E. C. Case, H. H. Rasche (regional and urban geog.), H. L. Shantz. *Africa South of the Sahara:* E. P. Hanson (Liberia), Roderick Peattie (U. of So. Africa), G. T. Renner, Jr., Leo Waibel, Derwent Whittlesey. *Northern Africa:* R. Van V. Anderson (Algeria and Morocco), Jean Gottmann, J. R. Smith

AUSTRALIA AND NEW ZEALAND

Frances M. Earle, T. W. Vaughan. *Australia:* J. E. Collier (econ. geog., esp. grazing), Anne M. Goebel, Griffith Taylor, C. M. Zierer. *New Zealand:* A. K. Botts, A. H. Clark (physical and hist. geog., land utilization)

L.C., U. California (1), U. Missouri (gl) `

INDEX

NOTE. Numerals in roman type refer to pages; those in *italics*, to the serial numbers of items in the Bibliography.

Titles of serials, series, and sets are printed in *italics;* other titles, in roman; titles of articles in periodicals and of parts or chapters of books are indicated by quotation marks. Separate titles are not ordinarily included in the index where they would appear close to corresponding subject captions (e. g. the periodical *Africa* is not indexed by title, since this title would occur next to the subject caption "Africa"); or titles beginning with the words "Atlas" or "Bibliography" or their equivalents in foreign languages (e. g. "Atlante," "Bibliografía," "Bibliographie," etc.). References to the pages on which such works are cited, or to their serial numbers, may be readily found by consulting the appropriate subject entries.

The names of persons and institutions as listed in the Appendix, pp. 276–294, are not indexed.